REFINED TASTES

The Johns Hopkins University Studies in Historical and Political Science

120th Series (2002)

1. Wendy A. Woloson,
Refined Tastes: Sugar, Confectionery, and Consumers in Nineteenth-Century America

Refined Tastes

SUGAR, CONFECTIONERY, AND CONSUMERS IN NINETEENTH-CENTURY AMERICA

WENDY A. WOLOSON

The Johns Hopkins University Press
BALTIMORE AND LONDON

Printed in the United States of America on acid-free paper
9 8 7 6 5 4 3 2 1

The Johns Hopkins University Press
2715 North Charles Street
Baltimore, Maryland 21218-4363
www.press.jhu.edu

Library of Congress Cataloging-in-Publication Data

Woloson, Wendy A., 1964–
Refined tastes : sugar, confectionery, and consumers in
nineteenth-century America / Wendy A. Woloson.
p. cm.—(The Johns Hopkins studies in historical and
political science; ser. 120, no. 1)
Includes bibliographical references and index.
ISBN 0-8018-6876-9
1. Confectionery—History. 2. Sugar—History. I. Title.
TX783 .W65 2002
641.8′6—dc21 2001002427

A catalog record for this book is available from the British Library.

To Joan, the best mom in the world
(Mwa!)

And in memory of Mildred Essick
and Rachel Lowrie

CONTENTS

PREFACE AND ACKNOWLEDGMENTS

When I tell people the subject of this book, they invariably make reference to my presumed sugar-laden diet. But in reality I am not a sweet person; I prefer the savory. I crave cheese—the sharper the better—as others lust for dark chocolate fudge. I rarely save room for dessert, satiating myself at the meal's beginning rather than end. So my interest in confections comes not from the stomach but from the head and the heart, from the intersection of two parts of my life, the intellectual and the artistic.

Earlier research I conducted into nineteenth-century American domestic life, coupled with a near obsession with late nineteenth-century food innovations like powdered gelatins, brought me to sweets from one perspective. It struck me that Victorian Americans created a "dessert culture" for themselves, which permeated all classes. Aesthetic preferences for ebullient designs, odd juxtapositions, objects under glass, and faux finishes carried over to people's culinary habits. These habits I saw most eloquently expressed in the early recipes suggested by the makers of Jell-O, Knox, and other instant gelatins. Broadening my focus both chronologically and thematically, I delved into the origins of this ebullient dessert culture. I was, of course, led straight to sugar, a commodity that dessert culture celebrated and accorded greater cultural purpose.

I also became curious about the concomitant sensibility of sweetness in Victorian American culture, which manifested itself in everything from lacy valentines and frilly dresses to characterizations of girls and women. It seemed to me that certain qualities characterizing desserts were also linked to, or expressions of, femininity. And this connection between the animate and inanimate made me even more curious about how this confluence came about. Ideally, I was hoping to unpack some of the ambivalent meanings of sweets and sweetness that we continue to live with today.

My approach, which at times may seem more intuitive than academic, is highly informed by my formal training as an artist. Long before I knew about surreal Victorian desserts, I was creating versions of my own as prints and sculptures. Cakes, bonbons, brownies, sundaes, and the like continue to make occasional appearances in creative work. While not my sole focus, I do return to confections every so often, taking comfort in the familiarity of their forms as I know them intellectually and artistically. How I came to consider confections a worthy artistic subject in the first place remains a mystery. They have a deep-seated aesthetic appeal that is beyond both my understanding and my capacity to articulate.

∴ ∴ ∴

This book came into being only after a tremendous amount of personal and institutional assistance. At every juncture I encountered people who not only supported my project without reservation but also offered sound advice without balking at a book on candy. The brief moments it takes for you to read their names do not even begin to approximate the time and effort they have contributed on my behalf. They willingly gave themselves to me via this work, and I thank them for it.

Institutional support for earlier versions of this project helped create the foundations for the finished product. Generous grants from the Winterthur Museum and Library and the University of Pennsylvania provided the funds and time necessary to conduct the main parts of my research.

Assistance with the nuts and bolts has come not only from institutions as such but also from the individuals working within them. Professionals and laypeople alike took a topic otherwise so easily dismissed seriously. The interdisciplinary nature of my book is reflected in the many people to whom I owe thanks and their disparate backgrounds and affiliations. Christa Wilmanns-Wells and Murray Murphey, both of whom I worked with at the University of Pennsylvania, and Steven Lubar of the Smithsonian Institution gave me the personal and intellectual support I needed to both start and finish. Uncovering much of the concrete evidentiary material would not have been possible without enthusiastic assistance from reference librarians and archivists. Neville Thompson at Winterthur, Pamela Whitenack at the Hershey Community Archives, and Fath Davis Ruffins and Vanessa Broussard Simmons at the Smithsonian Archives helped me gain access to collections little used, uncatalogued, and otherwise difficult to consult.

I am especially grateful to the readers who took such special care in perusing and commenting on versions of drafts. I took their feedback seriously and can confidently say that the more thoughtful parts of this book I owe to them. Paul Erickson, Martin Higgins, and David Miller—stalwarts all—read the manuscript in its entirety and provided invaluable critiques and essential support. Harold Screen took the time to give me generous feedback on specific parts of the manuscript and also provided me with important illustrations that I would not otherwise have had access to. I would also like to thank my colleagues at the Library Company of Philadelphia, including Jim Green and Phil Lapsansky, for bringing new sources to my attention, and to Jenny Ambrose, Sarah Weatherwax, Erika Piola, and Valerie Miller in particular for helping me obtain many of the reproductions used in the figures.

Other people's input is less easily quantifiable but was no less essential. David Miller good-naturedly competed with this book for my attention and patiently answered a constantly flowing stream of questions on topics ranging from nineteenth-century French vernacular to twentieth-century sports heroes. Bob Brugger, Melody Herr, and, especially, Janet Theophano urged me on at the finish line, providing emotional support just when I needed it the most. Bob Zecker, Terry Snyder, Linda Stanley, Mary Anne Hines, Bruce Compton, Paul Erickson, Martin Higgins, Anja-Maiike Green, Melissa Kerr, Ted Hobgood, Dave Jacobson, Donald Stilwell, and JoAnn Stilwell have remained my good friends through this entire project. They will be happier than I to see this book in print, if only so that its pages are off my dinner table and out of my thoughts. These people have been my strongest supporters. They reminded me that there is life beyond the computer screen, as did my canine companions, Elvis and Dean-O, who did not bring me hot cups of tea at midnight, did not pay the bills in my absence, and did not take themselves out for walks in order to make this happen.

Acknowledgments to family members always come last because, like sugar, they work their magic invisibly, making life sweeter especially when we are not aware of it. Thanks to my mother Joan, who eats Godiva chocolates in the tiniest nibbles I've ever seen; to my father Kent, who can keep that last Popsicle fragment on both sticks until the final bite; and to my brother Blake, who eats chocolate marshmallow ice cream melted, like soup.

REFINED TASTES

Introduction

Refining Tastes

The familiar nineteenth-century phrase "Home Sweet Home" nostalgically suggests a time of domestic bliss when innocence and harmony were embroidered into the fabric of family life.[1] In this postfeminist era, it sounds old-fashioned and possibly offensive to say and think that girls are made of "sugar and spice and everything nice," yet the language of sweetness, although occasionally applied to men to describe certain feats of athletic grace and aptitude, remains almost exclusively used to describe women and children. It seems natural to us to associate sweet things with femininity and innocence, even though this has not always been the case.

Until well into the nineteenth century, people held refined sugar in high esteem and ate it only in tiny amounts, because it was so expensive and hard to come by. The American elite often enjoyed sweet confections as status symbols at impressive dinners, sweets either of their own making or purchased from urban fruiterers and confectioners who specialized in provisioning the wealthy, whom they supplied with goods imported from England and France. In 1766, for example, Samuel Frauncis offered his Philadelphia patrons jellies, syllabubs, creams, custards, and a "universal Assortment of Sweetmeats, viz. Grapes, Grape marmlet, Cherries, Currant Jellies, Strawberries, Raspberries, yellow and green Peaches, . . . and a vast Number of other Things."[2] Among others, George Washington employed his own French confectioner, and Dolley Madison served ice cream in the White House. Copies of Hannah Glasse's *Complete Confectioner* and Elizabeth Price's book of the same title, published in 1770 and 1785 respectively, shared space in the libraries of America's cultural elite with works such as Blackstone's *Commentaries on the Laws of England,* Hume's *History of England,* and Chesterfield's *Letters.*[3] Even Thomas Jefferson's receipt book contained a recipe for ice cream, the result of his own personal experimentation.

Today, psychologists blame sugar for Attention Deficit Disorder, nutritionists consider it a dietary nuisance, and parents appeal to grocery store owners for "no candy" aisles. Although people of all ages and classes continue to desire, if not crave, sugar, they do not necessarily respect it, given its abundance. It is hard to believe that little over two hundred years ago, American consumers treasured sugar and the sweet things made from it. In the eighteenth and nineteenth centuries, sugar transformed even the humblest substance into a special treat for people who lived on a bland diet largely devoid of sweet things. The treasures in early shops dazzled the eyes and palates of the wealthy who could afford to enter. For example, in 1775, the confectioners James and Patrick Wright offered:

> Fine Carraway comfits, almond comfits, musk plumb comfits, band-string comfits, crispt almonds, Coriander comfits, almond chips, barley sugar, white sugar candy. . . . Fine almond biscuit, plain almond biscuit, spunge biscuit, fine broad biscuit, diet loaves, seed cake, plumb cake, liberty drops, common biscuit, pound cake, fine ginge cake, fine ginge nuts, queen's cake. . . . Cinnamon tablet, lemon tablet, orange tablet, ginger tablet. . . . Loaf, lump, and muscovado sugars, coffee, chocolate, West India and Philadelphia rum, cordials, &c., &c.[4]

Something more interesting happened to sugar in America than merely its democratization. Refined sugar entered American culture as a highly desirable product and therefore became a fitting repository of sentiment, a subject of consumer fantasies, and a signifier of social status. With the exception of a few other foodstuffs, like flour, whose shift from whole grain to bleached and refined signaled its perceived purification, most raw materials remained generic marketplace commodities and did not become such overt projections and subjects of consumer desire. During the nineteenth century, people not only made willing accommodations for sugar in their diets but also endowed it and things made from it with resonant cultural meanings. What is more, these meanings changed as refined sugar shifted from a rare and precious good to a prolific necessity.

Significantly, sugar's qualities of sweetness became linked with specific human traits. As a rarity, sugar signified male economic prowess in America. As it became cheaper and more prolific, however, it became linked with femininity: its economic devaluation coincided with its cultural demotion. Indeed, by the end of the nineteenth century, consumer and consumed had become entirely conflated: sweets had been feminized, and women were sweet.

Focusing specifically on the confectionery made from refined sugar enables us to see, at least in this one instance, how new commodities made their way into American culture and how people found meaning in them. This process raises important issues, addressed throughout this book. Looking at the various forms of confectionery, we can see how commodities assumed not only functional but also symbolic meanings, and the degree to which consumers and producers shaped those meanings. Advertising played a large role in animating commodities, at once influencing and articulating the meanings consumers derived from material goods. In the case of confections, advertisers blotted out the unpalatable by, for example, erasing any indications that sugar was produced by slave labor. When advantageous, they fetishized the producers of other goods, for example by portraying cocoa processors as exotic, bare-breasted native women. Producers and advertisers were able to link abstract qualities with the actual physical characteristics of sugar and its decreasing price, and consumers then came to link this devaluation specifically with feminization. There were larger cultural implications to this devaluation process, which shed light on the overarching relationships of people to their material universe.

As a "raw material," refined sugar was potentially many things, from invisible flavoring in tea to the basis for modest gifts of homemade fudge to flamboyant sculptures made of sugar paste. Unlike other marketplace goods, sugar's physical properties and versatile flavor excited the senses, especially the tactile, visual, and gustatory senses. It often appealed to all of these simultaneously, making it a source of pleasure and fascination. Americans' fairly rapid integration of refined sugar into their diets, however, was not just due to the alluring sensory attributes of confectionery. Equally important, sweet things also served as expedient repositories of sentiments, which could be triggered by the physical properties of confections and their packaging—handmade chocolates in heart-shaped boxes, for example. And while sugar's historical mystique up to the nineteenth century certainly flavored and colored its general reception in America, by the end of that century, its meanings had multiplied exponentially. Depending on the product and context, confectionery came to signify everything from romantic to familial love, from passion to piety. With few exceptions, the taste of sugar and the emotions associated with sweets made eating them a pleasure to think about and to do, even if it evoked the guilt of gluttony, promoted tooth decay, aggravated certain skin conditions, and invited the criticism of reformers. Consumers by and large enjoyed, anticipated, and celebrated the consumption of confectionery.

Consumers perceived retail goods, including confections, as objects with tangible, material properties. That goods also appeared in the marketplace mysteriously, almost magically, made them seem to transcend materiality, encouraging fantasy and desire.[5] Rather than assimilating goods into the human environment, people actually assimilated themselves into this new world of marvels brought to them by rapid industrialization. Products imitated the characteristics and abilities of people, and advertisers increasingly endowed inanimate goods with personal histories—their own biographies. Consumers' intimate identification with marketplace goods only made these commercial products more alluring. Purchasers took comfort in and with their newly bought items. They used them as mirrors and as social indicators. They employed them as signs to describe themselves and convey those constructed selves to others. This applied to confectionery as well, which is important to consider, given that sugar's complex system of meanings, evolved in the nineteenth century, still sticks with us today.

Sweet Dreams Become Realities

Sidney Mintz's book *Sweetness and Power* details the impact of refined sugar on British culture from the sixteenth century through the early nineteenth, a time period when it most signified power and status. Starting where Mintz leaves off, when sugar became fully assimilated into modern culture, I concentrate here specifically on America. Many factors contributed to refined sugar's democratization in nineteenth-century America, nearly a hundred years after such democratization had occurred in Britain. Increased international production and imports from foreign sources made sugar more plentiful in the American marketplace, while the falling price made it more affordable to average consumers. The Sugar Trust, an organization of the eight major refining companies formed in 1887, consolidated their twenty refineries into ten to streamline and increase production. Sugar had become so cheap by the end of the century that the Sugar Trust often kept production low so as to not flood the market. For example, by 1888, the Sugar Trust had a production capacity of 34,000 barrels a day but produced little more than 23,000.[6]

Technological improvements aided these refiners. New steam-powered equipment, which reduced the time needed to refine a batch of sugar from two weeks to twenty-four hours, included a granulating machine, the vacuum pan, and a centrifuge, introduced in 1848, 1855, and 1860 respectively.[7] Before these innovations, laborers had performed these processes by hand, cooking down sugar syrup in kettles over an open flame and relying on gradual cooling to granulate and evaporation to refine the sugar. The growth of the sugar industry inspired other improvements in sugar production technology as well. According to government records, people applied for 79 sugar-related patents in the 1850s, and the figure jumped to 257 during the 1860s—a threefold increase.

Technological innovations and price supports such as favorable tariffs for domestic sugar refiners helped provide Americans with the sugar they increasingly craved. American sugar consumption rose gradually during the eighteenth century and exploded late in the nineteenth century. Statistics vary, but all sources agree that Americans' sugar consumption increased exponentially during the nineteenth century (see fig. 1). Domestic beet sugar production was perfected in the late 1870s and, along with beet sugar from Europe and cane sugar from Cuba and the Sandwich Islands (as Hawaii was then called), where the United States was increasingly involved, contributed to supplying the now mature American sweet tooth.[8] By the early 1870s, the average American consumed almost 41 pounds of (mostly imported) sugar a year,[9] over six times what his or her counterpart had eaten in the 1790s.[10] Americans not only devoured refined sugar, they also made things with it, and by midcentury the United States had developed a flourishing confectionery trade, which became a leading exporter by the century's end. Whereas fifty years earlier Americans

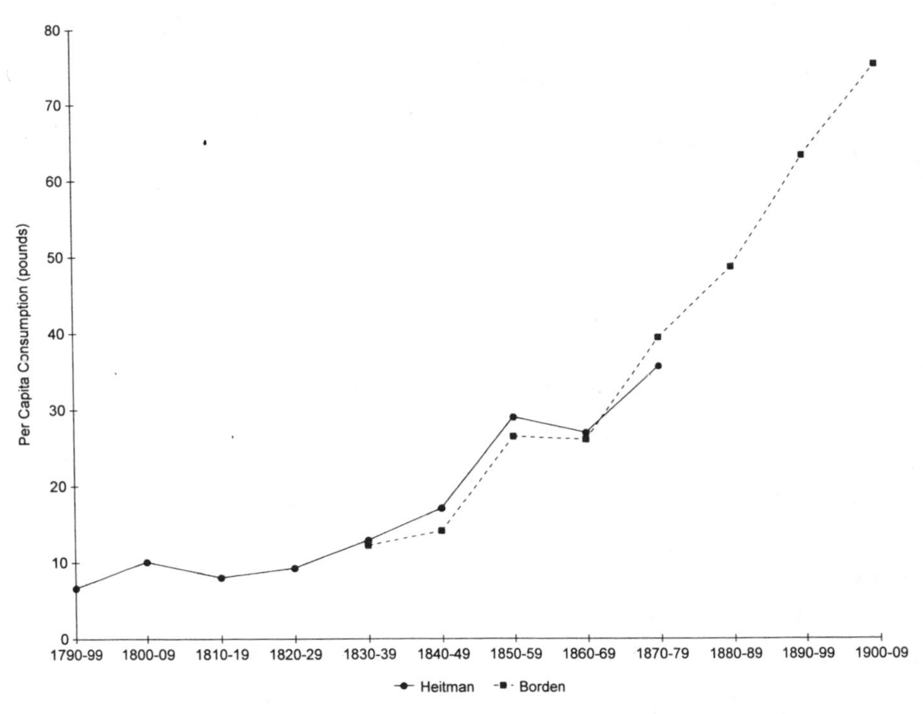

FIGURE 1. U.S. per capita sugar consumption rates, by decade. Neil Borden, *The Economic Effects of Advertising* (Chicago: Richard D. Irwin, 1942), 280; John Heitman, *The Modernization of the Louisiana Sugar Industry, 1830–1910* (Baton Rouge: Louisiana State University Press, 1987), 58.

had eaten imported confectionery almost exclusively, by the second half of the nineteenth century, American confectioners produced so much that they successfully sold their goods in overseas markets: in 1865, the export trade in confectionery was valued at $26,429; by 1881, it had grown to $73,253; by 1890, to $179,276; and by 1895, with the help of improvements in confectionery manufacturing technology and beet sugar production, to $712,552.[11] Not only that, but Americans continued voraciously to consume what they did not export. By the end of the nineteenth century, the average American wallowed in refined sugar, consuming pounds and pounds of it in an ever-increasing variety of forms—an average of 68 pounds per capita annually by 1901.[12]

If producers successfully delivered sugar to the marketplace, consumers put it in their mouths. Confectioners offered refined sugar to consumers in cultural forms that allowed purchasers to endow them with meaning.[13] Working with sugar became the livelihood of more and more people. For example, the Philadelphia market in 1820 supported one confectioner for every 6,854 people. By 1860, the ratio was one confectioner for every 867.[14] Indeed, just about anyone who wanted to do so cashed in on the market for sugar. Successful confectionery equipment companies like Thomas Mills and Valentine Clad in Philadelphia supplied candy-making machinery to both large and small businesses. Patent records reveal the efforts of entrepreneurs who exploited inventions ranging from improvements in sugar refining to confectionery-making equipment and candy recipes.

Social factors both encouraged people to ingest sugar and circumscribed its use. These included the influence of popular literature, industrial expositions, and advertising, which imbued refined sugar products with recognizable qualities. As sugar made its way into the American diet, it also occupied more space as a symbolic entity in people's minds. In 1908, Paul Vogt stated matter-of-factly that "refined sugar has become a necessity in every household," although a hundred years earlier, people intent on equalization had lobbied for a tax on refined sugar because it would most affect the wealthy. Sugar had become so essential in people's lives that consumers opted for poorer, sometimes adulterated, grades rather than go without when prices reached beyond what they could afford.[15] The decisions that nineteenth-century Americans made regarding sugar consumption are important not only because they shed light on our ancestors' material life but also because they helped establish the modern American dependency on sweeteners.

In addition to that of sugarcane, the political economies of cacao and ice contribute to the story of confectionery, because these often served as the vehicles that carried refined sugar to people's mouths.[16] But by looking at confections themselves, the "cultural products" of these raw materials combined, it is evident that consumers' relationships with these commodities changed both materially and conceptually. People were not satisfied merely with dissolving sugar into other foodstuffs; they wanted to see the sweetness as they were eating it—as hard candy, as ice cream, as bonbons, as festal cakes. Sweets in all their forms not only occupied a greater physical presence in the cultural landscape but contributed to a vibrant material language. People conveyed social and political ideologies more effectively through things than through words and often enlisted ephemeral and seemingly trivial objects, like confectionery, to do so.

⁖ *Sweetness and Hype* ⁖

The visual advertising that matured during the nineteenth century conveyed much of this information. Merchants had once relied on signs to call attention to their stores, but these had given way by the mid nineteenth century to advertising claims that created "impressions" about products—via text and image—instead of merely reciting basic facts. Advertising transformed plodding lists of newly available dry goods into vehicles that matched ethereal, abstract, and often regenerative traits to individual products. People gave meanings to these commodities by melding the goods' actual physical traits with perceptions about them as defined, clarified, and explicated by advertising images, which consumers in turn selected, accepted, and implemented as social currency. Much to the chagrin of contemporary cultural critics, advertisers used an increasing number of strategies and contexts to disseminate their messages. Ads escaped from the margins of newspaper columns and appeared in the form of leaflets, pamphlets, billboards, souvenirs, letterheads, packaging, premiums, and myriad other objects and ephemera that made claims via eye-catching typefaces and tantalizing images. Larger institutions like industrial expositions, which displayed working machinery and sold the products the machines generated, were really just grand forms of advertising touting the achievements of companies and nations. Everything from text-based enticements to window displays tried to be artistic, appealing to the eye rather than the intellect.

By the end of the century, the proliferation and increasing impact of such visual images were undeniable and inescapable. According to Thomas Schlereth, American advertisers spent $95 million on their promotions in 1900, ten times what they had spent in 1865, even though the population was only about two and a half times greater.[17] For everything from umbrellas to candy, advertising worked to associate specific products with identifiable images, which collectively reinforced one another, creating an overarching sensibility that influenced consumers by its sheer omnipresence. Late nineteenth-century Americans were not passive recipients of these new advertising strategies. In fact, the flood of goods and fierce competition brought about by a maturing industrialization forced Americans to be that much more savvy about their purchasing decisions, and they did not allow themselves to be bowled over by the latest advertising campaigns. Active consumers even as children, they made conscious decisions about what they bought and how they used what they bought.[18] But the push of advertising, the chief mode of communication between producers and consumers, radically altered the visual landscape, especially in urban areas.

Confectioners, like other businessmen, took advantage of the new trend in visual aesthetics, incorporating representational imagery into print ads, window displays, and some of the most interesting trade cards circulated. They capitalized on the attractive qualities of their products, using them to entice customers. In response, people became more sophisticated at "reading" pictures and goods. Honed visual skills enabled consumers to decipher the visual vocabulary used by advertisers more expertly. Yet consumers' preferences often determined which advertising strategies producers implemented to begin with. Moreover, consumers only bothered to decipher the meanings of images deemed worthy, easily screening out the unimportant or uninteresting, much as we do today. These same critical sensibilities informed people's interpretations of three-dimensional objects and inculcated attitudes supportive of existing class hierarchies. "The making of nineteenth-century class society was not only about transformations in the relations of people to the means of production but also about their massively changing relations to systems of commodity exchange and styles of consumption," Victoria de Grazia observes.[19]

Confectionery provided one such example of this process. People judged sweets initially by what they looked like, and they assumed many different physical forms, alive with symbolic possibilities and geared toward attracting specific consuming audiences. Food in general can be viewed as both "a highly condensed social fact" and a "marvelously plastic kind of collective representation," the anthropologist Arjun Appadurai remarks,[20] and this applies to confectionery in particular. Through changes in temperature, flavorings, ingredients, handling, and packaging, refined sugar (often combined with chocolate and ice) assumed many guises. It was crystallized into hard, crunchy rock candy, churned and frozen as ice cream, whipped with butter and made pliant for cake frosting, and molded into soft, creamy bonbons.

The same material versatility was not shared by molasses and rum, other products derived from sugarcane. While they both played significant roles in the early American economy, people did not incorporate these commodities into their emotional lives in quite the same way they did refined sugar and confectionery. Molasses, a by-product of the sugar-refining process, found many uses as a cheaper sugar substitute and as a basis for manufacturing alcoholic beverages, including beer and, more economically and socially significant, rum. "There can be little argument that rum helped in the conquest of North America," writes John McCusker, who notes that molasses was not only the "poor man's substitute" for sugar and barley for brewing but also most often used commercially in the distillation of rum.[21] Rum also provided a significant source of wealth for the early colonists. However, consumers did not associate rum, a by-product, with its original source—sugar—and for that reason it is not integral to this story. Molasses, on the other hand, remains an instructive counterpoint to refined sugar because of what it represented, especially when it came to class issues. Literally the dregs of refined sugar, molasses found its way to the tables of the less fortunate; the two commodities clearly marked class disparities and, toward the end of the nineteenth century, racial differences as well—dark viscous liquid versus white granules, impure versus pure, unrefined versus refined.

Sugar did not play a neutral role in American culture: it garnered specific cultural connotations from associations based on its appearance, packaging, concrete economic value, and its frequent comparison to people. Photographs, cookbook illustrations, trade journal diagrams, and eyewitness descriptions of the time evince how producers intended different forms of sugar to appeal to different groups of consumers. Sugar did not remain just a reticent flavoring for tea in American diets. It was incorporated into cakes and fudge, dissolved into sauces, sprinkled on fruits and vegetables, enlisted as the foundation for paste that made sculptures, used as a basis for bonbons, and included as an essential component of ice cream. More important, sugar *as* these various forms assumed widely shared cultural meanings. By the end of the nineteenth century, it had lost its original meaning as a sign of masculine power and had been endowed with properties perceived as feminine, such as refinement, gentility, piety, and weakness. Contemporary sugar semiology extended into packaging and consumption, too: heart-shaped boxes of chocolates communicated romantic intentions, three-tiered, white flowered cakes were synonymous with brides, ice cream saloons emerged as women's domains, and homemade fudge became a sign of domestic bonding.

Simply put, refined sugar in these various forms, often mated with chocolate and ice, embodied and communicated meanings. But these meanings were not random; rather, by becoming part of a coherent and consistent language that linked sugar with qualities of class, gender, and age, they reinforced the existing social hierarchies. Popular sources from the nineteenth century document the progression of this process and demonstrate how, through an utter saturation of references to sweetness and refinement—mostly in sources outside the sugar industry—both men and women came to accept these associations as not only logical but also as "natural," part of their "habitus."[22] What is more, our present-day meanings of confectionery emerged from those developed in the nineteenth-century American market. Rather than escaping them, we have built upon them.

Even Americans' daily lexicon accommodated the increasing role of sugar in the culture. Before the 1850s, "sweet" as a quality described the fragrance of fresh flowers; when used to characterize a woman, it referred to her general mien or to matters of hygiene—the smell of her breath or condition of her skin, for example. The material characteristics of sugar and confections themselves, coupled with an earlier history of "sweet" things (not associated with sugar) being particularly gentle and feminine, reinforced the later connections made between confections, women, and children. The language of sweetness is so familiar and natural to us now that we take it for granted: girls are "sugar and spice and everything nice," and we take refuge in the safety of our "home sweet home." But a century and a half ago, these references were only beginning to have cultural resonance.[23]

Sugar's conversion from masculine to feminine, of course, occurred within and was influenced by a larger cultural context. For example, elaborately packaged chocolates and bonbons could never have been commercially viable without the concomitant rise in the celebration and sentimentalization of holidays like Valentine's Day, Mother's Day, Easter, and Christmas. The popularization of wedding cakes would not have happened without a corresponding shift to a display ethic that involved genteel women especially, not only as the chief consumers but also as the main showcases of that consumption—ornamental vehicles for displaying economic class and social status. In turn, femininity was readily translated into fancy sugar work. Homemade confections would not have met with widespread acceptance without the improvement of printing and distribution systems that made domestic confectionery manuals so pervasive, a labor system that allowed genteel women to produce luxury items in the home, and a sophisticated manufacturing system that made other sweeteners, such as honey, maple sugar, and molasses, seem quaint in comparison. Aiding sugar's feminization was the widely held assumption that as a nonessential good with little nutritional value, sugar most likely appealed to the cravings of women and children—groups of the population assumed to have little control over their mental and physical impulses, which led them to desire things neither necessary nor practical. In nineteenth-century America, the conflation of relative economic empowerment and social empowerment transmuted sugar, once a highly prized substance, into an ephemeral material used to make things that were sweet, delicate, refined, and feminine.

An age component figured in this as well. Although consumers of all shapes, sizes, and backgrounds ate all sorts of candies, they adhered to a general pattern of consumption that linked their life stages with the kinds of confectionery they preferred. Rather than transgress age-related boundaries for more affordable sweets, people generally chose a cheaper brand of the same generic type of candy. For example, an indigent young woman would opt for inferior bonbons rather than penny candies, because bonbons in taste and texture appealed to her mature palate. People did not always clearly articulate these categorical divisions, nor were they always conscious of the commodity-consumer relationship. However, these general consumption categories created a schema within which people ordered, ranked, and understood commodities, which resembled the social categories they used to order, define, and rank one another.

Brightly colored, hard, cheap penny candies were the first candies mass-produced in America, quickly becoming the desired treats for children, who learned the value of money and the cultural cachet of possession through selecting, buying, trading, playing with, and finally eating their penny candies. Ice cream mediated social relationships in ice cream parlors and other public settings, much like its adolescent consumers. Soft-centered candies like chocolates and flavored bonbons became confections designated for older consumers and the sexually maturing who appreciated the subtle and sensuous pleasures of savoring something slowly. Costlier and more sumptuous candies, they pleased the more sophisticated palates of teenaged and adult women. Bonbons, especially chocolate ones, also became part of lovers' vocabulary—purchased, given, and eaten by them during the blossoming of courtship. These tokens expressed romantic sentiments materially rather than verbally and initiated an obligatory cycle of gift-giving. "The young man who has

FIGURE 2. Red Cross Cough Drops. *Confectioners' Journal,* January 1890, 25. Library of Congress.

spent many pleasant moments as a boy with nose pressed to the candy counter, seems to know by instinct that his surest weapon as a suitor is a box of candy," a contemporary remarked.[24] A specific form of confectionery also marked the culmination of this romantic love ideal in the form of the ceremonial wedding cake, which, as the symbol of the bride, facilitated her proper integration into a larger kinship network upon marriage. Finally, sugar as comfortably domesticated and aging appeared in the form of homemade fudges and horehound candies, which family members often gave to each other as a gesture expressing warmth and goodwill, referencing old-fashioned quaintness and wholesomeness.

Sugar's economic depreciation coincided with its cultural devaluation and reemergence as a "feminine" substance. But even as a socially demoted prod-

FIGURE 3. Because molasses was often associated with the south, advertising for molasses candies frequently incorporated racist imagery. *Confectioners' Journal,* July 1899, 28. Courtesy of Harold and Joyce Screen.

uct, sugar and the confections made from it adhered to a familiar hierarchy based on appearance that equated purity and gentility. Therefore, sweet things were designated for those with refined palates; eating such delicacies made consumers even sweeter and more refined. Advertisers picked up on and promoted these associations, which had for years appeared in forms of popular literature like medical treatises and etiquette books. Using men rather

FIGURE 4. Sweeter confections enjoyed daintier purveyors in the language of confectionery advertising. Philadelphia Candy Store. Trade card, ca. 1890. Warshaw Collection of Business Americana. Archives Center, National Museum of American History, Smithsonian Institution.

than women to promote confections that were less sweet or that had medicinal purposes was a common turn-of-the-century advertising strategy (fig. 2). Gap-toothed yokels or monkeylike figures, replete with racist implications, hawked less sweet molasses candies and licorice (fig. 3). Angels, however, delivered superior saccharine delights (fig. 4). Meanings had become so ingrained as to seem natural—Roland Barthes's success of the sign and Karl Marx's success of commodity fetishism. Marx analogized this to religion, writing of the marketplace, "There the products of the human brain appear as autonomous figures endowed with a life of their own, which enter into relations both with each other and with the human race."[25]

If refined sugar, cocoa, and ice in the form of confectionery created sets of meanings for consumers individually and collectively, how did a consensus of meaning come about, and to what end? And if these meanings were inextricably linked with the people who consumed certain kinds of confections, then what can this tell us about people's changing conceptions of animate and inanimate things? Looking at the democratization of sugar through popular forms of confectionery in the nineteenth century reveals the creative and dynamic ways in which people incorporated goods into their lives, treating them almost as entities with lives of their own. Individuals chose to use sugar and to eat confections, just as they chose to create and accept their collateral symbolism. As we shall see, this cacophony of voices, from consumers, temperance workers, and domestic scientists to doctors, advertisers, and abolitionists, weighed in on the sugar question, contributing to a larger conversation about just how sweetness was to infiltrate the culture. Americans arrived at a coherent and consistent structure of meanings derived from both traditional and newly forming ideas circulating about the nature of sweetness, of confectionery, and of themselves.

CHAPTER ONE

Sugarcoating History

∴ *The Rise of Sweets* ∴

Evan Morgan, a displaced Londoner who made corset stays and children's coats after moving to Philadelphia, began advertising the goods sold in his shop in 1730. Along with his own tailored creations, cheap whalebone, and stay trimming, Morgan sold "very good Chocolate, Wine, Rum, Melasses, Sugar," and "several other Sorts of Goods," all "very reasonable."[1] Today we are used to seeing these foodstuffs in stores, but they were remarkable sights for eighteenth-century Americans. Given where they originated and what it took to procure them, it was quite unusual for these goods, especially chocolate and sugar, to share the same shelf space in an American shop at that time.

It was not for another hundred years that refined sugar, chocolate, and the confectionery made from them began their transformation into common staples. The raw materials used to make confections had long histories, which influenced how nineteenth-century consumers perceived them as marketplace goods and gave them meaning as cultural objects. The complex apparatus that delivered these histories and meanings to consumers did so in a variety of ways: through established traditions that had become ingrained cultural mores; through a political awareness about contemporary cultivation and production realities that motivated groups such as abolitionists and reformers;

through a general knowledge of other more prevalent and convenient sweet substances, such as honey and maple sugar; and through the rhetoric and reputations of eighteenth- and nineteenth-century sugar traders and politicians.

Many of these histories proved so enduring that they colored people's conceptions of refined sugar generally and confections particularly. Some elements of these histories faded from view, others remained, and yet other, new histories accreted to form meanings that people came to associate with sugar and its attendant products. Sugar gave confections their raison d'être. Like the other constituent ingredients of sweets—including ice and cacao beans—sugar possessed social meanings and ritual contexts much different and far removed from the feminized, ephemeral treats they became in nineteenth-century America.

The widely divergent geographical and chronological trajectories of confectionery's three key ingredients—cane sugar, cacao beans, and ice—were trails of contestation, conquest, and ownership. In addition, the sometimes purposeful and sometimes serendipitous ways in which users altered and combined these raw materials demonstrated the overwhelming human desire not only to consume them but also to make significant objects of them. These raw materials were the subjects of many experiments in cultivation and also became the stuff of aggressive trading: all three natural materials, as valuable commodities, traveled across and between various continents in an attempt to overcome climatic barriers while following human appetites—often with tremendous, unstinting effort. They all required intensive manual labor for their growth, cultivation, and processing; sugar and cacao cultivators even implemented and perfected the system of slave labor later adopted by America's cotton kings, making the goods undeniable manifestations of the masculine power that literally embodied human domination. Plantation owners, traders, merchants, and wealthy consumers—all who came into contact with these commodities—dealt in the conquest not only of land and crops but also of people. That sugarcane, cacao beans, and ice could only be produced and harvested in discrete, limited areas of the world made them sought-after items available only to very wealthy consumers, who used them as substances in rituals, commodities for trade, and domestic medicines.

People devoted an inordinate amount of energy to improving the manufacture and distribution of sugar and cacao in particular. They organized production trusts and merchant monopolies. They harnessed steam power and devised more reliable shipping strategies. They experimented with alternative agricultural sources and instituted favorable trade restrictions and protective tariffs. Most of these developments gained momentum in the seventeenth century, when confections and their constituent raw materials made their way into the marketplaces and onto the dinner tables of Europeans and colonial Americans.

Sugar

It is impossible to determine where and when, precisely, cane sugar originated. Sidney Mintz and his sources place the domestication of sugarcane in New Guinea, "and very anciently," around 8000 B.C., and believe that it had reached the Philippines and India from there by 6000 B.C., but whether people actually extracted sugar from the plant then remains uncertain.[2] Others think it came from China, some 2,000 years before it reached Europe in A.D. 1000.[3] Mintz places references to the actual production of sugar from cane "well into the Christian era," however, although perhaps earlier in India; his sources are unclear. Other sweeteners, like honey, predated cane sugar and suggest a perpetual human sweet tooth. But cane sugar's versatile material properties (it could disappear into a cup of hot tea or be molded into a tower), and exotic nature (it came from the remote tropics) made it even more seductive to early producers and consumers than goods produced and traded domestically in Europe. Cane sugar eventually overtook other sweeteners, traveling to Arabia, Sicily, Cyprus, Malta, Morocco, Spain, and North Africa.[4] Interestingly, it was at about this same time, roughly A.D. 700, that the T'ang rulers of China began to use ice as a food preservative and for summer air conditioning, and only a half a century before the Mayans' first depiction of a liquid chocolate drink.[5] But because of the distinct growing conditions for these natural supplies and the relative immobility of world civilizations at the time, use and cultivation necessarily remained confined to the people of discrete geographical regions.

Northern Europeans' relationship with sugar began around A.D. 1000, when the Spanish began using it for medicinal purposes. A hundred years later, journals from the court of Henry II of England recorded sugar usage but indicate only small amounts; at this time, Europeans who used sugar considered it one among many spices that added complementary flavors to culinary dishes—unlike the way we think of it now, as a key and elemental flavoring. Sugar cultivation and processing entered into Mediterranean and European cultures via the Crusades, and Europeans started growing sugarcane near Tyre and on Crete and Cyprus. From A.D. 1200 on, Antwerp reigned as the refining center for European sugar, followed later by other port cities such as Bristol, Bordeaux, and London. About this time, powerful men counted sugar among their material displays of power. Mintz offers the example of Caliph al-Zahir in the eleventh century, who incorporated some 73,300 kilos of sugar into his post-Ramadan celebrations, presenting "table-sized" sugar works of figures and palaces—a material statement made even more emphatic because of contemporaneous famine and plague. Soon afterward, this practice of showy display spread to the courts of England and France. Rulers there, too, articulated their power in immediate and incontrovertible ways. In 1287 alone, Edward I ordered a total of 1,877 pounds of sugar for his use; the next year, the court's sugar intake totaled 6,258 pounds—not even close to the Caliph al-Zahir's 161,597 pounds but impressive nonetheless, especially when one considers that this was over three centuries before the wealthy could afford to eat sugar, even by the spoonful.[6] Edward's court probably ate rose and violet sugars at the end of meals, "as a medicine to comfort the full stomach," Joop Witteveen suggests.[7] The rest of the king's sugar did not all disappear into "made dishes," but his cooks transformed it into material manifestations of the ruler's power via ornamental sugar work. Sugar sculptures, or "sotelties" (also called "subtleties"), played a crucial role in court life by embodying the king's power and presenting it to his guests in an incontrovertible way. "By eating these strange symbols of his power, his guests validated that power," Mintz explains.[8] Guests not only validated the ruler's power but also literally embodied it by ingesting parts of these imposing sugar works. Consumption in this context did not mean merely using up precious sugar; it was also a way in which people established and maintained political and cultural hierarchies.

The continued spread of sugarcane cultivation and refining made sugar more accessible to more Europeans. The nineteenth-century writer John Scoffern credits the Crusades with having made sugar a coveted good among Europeans: "The sturdy warriors of the cross, on their return to the west, began to [desire] many oriental luxuries for which they had acquired a taste. An oriental commerce was speedily established, and Venice became the great emporium of the riches of the east:—Of these sugar was one."[9]

It was not, however, until the 1300s and 1400s (when Spain and Portugal experimented with sugar production in their colonies on the Canary Islands and São Tomé) that sugar moved beyond the European courts and infiltrated elite households. Columbus first brought sugar to the New World from the Spanish colonies in 1493. By the sixteenth century, entrepreneurs had turned their efforts to making profits from the growing collective European sweet tooth, a difficult task given sugarcane's fragile nature. (The plant, very sensitive, requires constant watering, a long growing season—sometimes over twelve months—and a tropical climate. Transporting the cane from plantation colonies to European refining centers also proved perilous to the delicate substance.) By 1520, the Spanish had introduced sugarcane to Mesoamerica, where it joined cacao for perhaps the first time. And by 1526, Brazil had begun regular shipments of sugar to Portugal, inaugurating what Mintz calls "the Brazilian century for sugar." The spread of European colonization in the tropics increased the chances that these countries could grow their own cane rather than relying on expensive foreign imports. Caribbean plantations run by the Dutch, French, and British eventually surpassed them in the seventeenth century.[10]

Another practical factor limiting the production of cane sugar in the preindustrial world was the need to refine it—stalks of raw cane are not finished commodities. The grown stalks of cane are chopped down, hewn into segments, and fed through mills that extract the juice, which is then boiled down and filtered, finally yielding a granulated product. Having the capital to invest in labor and machinery to do all this determined a colonizing country's ability to supply its own domestic markets with sugar and export it for profit. By the 1650s, England accomplished this, importing and refining its own sugar for its own market rather than relying on that refined in Holland.[11] Through the centuries, as Britain's taste for sweet things increased, it produced and imported more refined sugar and exported less: overall consumption continued to rise, as did the entire amount produced and refined. By the late 1600s, sugar had become so prevalent in Europe that people valued it as a food in its own right; its former roles as a spice, a condiment, and a medicine began to wane.

Indicative of the colonists' relative disempowerment in the world market at the time, democratized use of sugar (and of chocolate, for that matter) occurred almost a century later in America than in Europe. The English introduced sugarcane to Jamestown, Virginia, in 1619, but the growing season proved too short; Virginia winters were too cold for the crop to flourish.[12]

The appetite of the European elite for sugar was voracious, and they consumed as much as they could afford and reexported what little remained to their New World colonies. Soon after Virginia's failed attempts to grow sugarcane, British Captain John Powell settled Barbados, in 1627; by 1655, Britain had solidified its position as the chief sugar producer, which it held for well over two hundred years, until the sugar beet, a hardier plant, was developed as an alternative to sugarcane.

By 1660, Britain enjoyed an almost total monopoly on the sugar industry and was able to export twice as much as it consumed.[13] Richard Blome detailed the British control of the sugar trade in his 1678 description of Barbados, which appeared in *A Description of the Island of Jamaica*. In it he described the abundance supported by the islands: "these *Commodities,* especially *Sugar, Indico, Cotten,* and *Ginger,* here are in such great abundance that about 200 sail of *Ships* and *Vessels,* both great and small, have yearly their loading; which after Imported in the several ports of *England* and *Ireland,* is again in great quantities exported to Foreign parts, to our great enrichment."[14] Still a highly prized good (in 1654 Britain, for example, a shilling bought two pounds of butter, yet only a pound of sugar if purchased in quantity),[15] sugar was the wonder substance. *The Queen's Closet Opened* of 1662, for example, contained recipes that used sugar as an essential preservative and medicine. Some recipes claimed to cure the palsy, help the digestion, or be a "Purge for Children or old men." Other instructions told how to candy flowers, make preserves, and even create rudimentary forms of hard candy, like the recipe, "To make Sugar of Wormwood, Mint, Anniseed, or any other of that kinde." Remarkably, *The Queen's Closet* even included instructions for making walnuts (complete with shells) and facsimile collops of bacon out of marzipan paste.[16]

Increasingly voracious appetites for sugar required even greater bodies of labor to produce it. From the late seventeenth century on, the Dutch and English, particularly, used ever-growing numbers of enslaved Africans to work their tropical sugar plantations. According to Sidney Mintz's figures, between 1701 and 1810, Barbados, a mere 166 square miles in area, received 252,000 slaves, and Jamaica, 4,244 square miles in area, received 662,400 slaves to work the plantations. This influx of labor fed unstinting British consumption, which increased by more than 400 percent during the century, from 4 to 18 pounds per capita per year. Britain imported more, consumed more, and exported less: in 1700, she imported 50,000 hogsheads of sugar and exported 18,000; in 1730, she imported 100,000 hogsheads and exported 18,000; in 1753, she imported 110,000 hogsheads and exported a meager 6,000.[17] It is no wonder that the sweet tooth of the American colonists took so much longer to develop, given the relatively small amount available to them on the market and the exorbitant prices they had to pay compared to staple goods, even during times of fluctuation. For example, during the 1720s, muscovado (raw sugar) imported from the West Indies sold wholesale from 23 to 40 shillings per 112 pounds on the Pennsylvania wholesale market, while the same amount of flour sold for between 9 and 12 shillings. In 1747, this figure for muscovado reached a high of 58 shillings, whereas flour never sold above 21 shillings (in 1748), and this was an anomaly.[18]

Significantly, however, high prices did not thwart elite Americans, whose sugar habits evolved alongside and in concert with their coffee and tea addictions—they enabled and provided excuses for one another. By the 1700s, the idea of sugar as a symbol of economic power and luxury had been firmly implanted in colonists' minds. Indeed, Americans exhibited many of the affectations of their British relatives and fellow citizens, including the penchant for lumps of sugar in their ceremonial tea. The desire for sweetmeats and other sweet delicacies (imported, of course) already existed, and it was reinforced among the rich, not only by fancy confectioners, but also through cookbooks imported from London publishers. Works like Hannah Glasse's *The Compleat Confectioner,* which first appeared in 1751, and Elizabeth Raffald's *The Experienced English Housekeeper,* first published in London in 1769, which both contained recipes for ice cream, pastries, cakes, and even sugar sculptures, went through many editions. Yet although recipes for confections appeared in printed form, this did not necessarily mean they were popular, for many recipe books, along with other printed materials of the time, were destined for the shelves of wealthy people who valued these books as much for their value as status objects themselves as for the information between their covers.

Other forms of print media, however, began to circulate ideas about sugar to a wider audience. For example, from its very first issue, the *American Weekly Mercury* published New York and Philadelphia market prices for sugar: in late 1719, muscovado sold for 40 to 45 shillings per hundredweight (112 pounds) and "mallosses" for 1 shilling, 6 pence per gallon.[19] The earliest sugar advertisement appearing in the *Pennsylvania Gazette* was Samuel Keimer's of 8 May

1729, which offered "very good choice Loaf Sugar" for 2 shillings per pound.[20] There was room in the American market and diet for a variety of types and qualities of sugar. For example, Peter Delage, a Philadelphia sugar baker, sold all kinds of sugars in 1736. He offered: "choice double refin'd Loaf Sugar at Eighteen Pence per Pound, and single refin'd Loaf Sugar at One Shilling per Pound, Powder [*sic*] Sugar, Muscovado Sugar suitable for Shop or Family use, Sugar Candy and Mollasses."[21] Yet it was not until well into the nineteenth century that the domestic market for confectionery—separate goods made with such sugars—became viable.

People on both sides of the Atlantic wanted to sweeten their diets, regardless of the costs or consequences. Tellingly, people heard little outcry about the exploitative methods used to produce sugar. Some Britons launched antislavery campaigns, which Americans later took up, but they experienced varying degrees of success. In *An Address to the People of Great Britain* of 1791, William Fox argued that consumer boycotts could alleviate the use of sugar and hence the need for slave labor. He figured that if a family using 5 pounds of sugar a week quit eating it, "with the proportion of rum," for twenty-one months, they would "prevent the slavery or murder of one fellow-creature." He further extrapolated that "eight such families in 19½ years, prevent the slavery or murder of 100, and 38,000 would totally prevent the Slave Trade to supply our islands."[22]

In a pamphlet entitled *No* RUM!*—No* SUGAR! published in 1792, the anonymous author penned an imaginary conversation between a slave, "Cushoo," and "Mr. English," resulting in the latter's decision to "leave off the use of Rum and Sugar, at least till the Slave Trade is abolished." Cushoo suggests to Mr. English that he abstain from eating sugar, and the latter asks, "But what are we to do without these articles?" Cushoo answers, "'Stead of Rum, drink brandy, Wine, or good strong Beeré." Mr. English thinks he can forgo rum, but giving up sugar seems nearly impossible:

E. Rum, I think I can pretty well dispense with; but what must I do for Sugar?

C. What dye do 'fore Sugar made, Massa?

E. Go without I suppose.

C. O!

E. But how must I drink my Tea and Coffee?

C. Drink tea as they do in China, Massa, and Coffee as dey do in Turkey—widout Sugar.
E. But sweet'ning is to some things absolutely necessary.
C. Honey very good sweet!
E. And a very dear one too!
C. So Sugar, Massa, very dear and like be more dear.
E. Sugar is sometimes wanted for medicine.
C. Den let de Pothecaries sell it; and dye no sell much, you no eat much.[23]

The English poet William Cowper expressed similar sentiments. In "Pity for Poor Africans" his subject was the hypocritical consumer whose guilt about slavery could not overcome his cupidity: "I pity them greatly, but I must be mum, / For how could we do without sugar and rum? / Especially sugar, so needful we see; / What, give up our desserts, our coffee & tea!"[24] Due to economic rather than humanitarian concerns, the British colonies ceased their importation of slaves in 1807 and finally abolished slave labor on their plantations from 1834 to 1838; the industry had by then found more economical sources of refined sugar and cheaper, more mechanized ways of producing it. Other countries abolished slavery on their sugar plantations in the ensuing years.

Consumer insensitivity to or naïveté about the larger political issues engendered by labor and production realities characterized sugar's presence in the marketplace throughout the eighteenth and nineteenth centuries. Notwithstanding the pleas of abolitionists, people on both sides of the Atlantic continued to consume sugar in ever-increasing quantities. The abolitionist newspaper the *Pennsylvania Freeman,* for example, carried many advertisements for stores selling "Free Groceries" (i.e., produced with free rather than slave labor) in the late 1830s. Robert M'Clure's was one among a number of enterprises that offered such staples, and his stock included "Double, Single, and Lump Sugar, Canton Sugar in bags and boxes; old and green Java Coffee, St. Domingo, Laguira, and Jamaica do. [ditto]; Eastern Island and N. York sugar-house Molasses; East India rice; Free Chocolate, made from St. Domingo Cocoa, &c."[25] These shops, however, were short-lived; regardless of conscience, people's desire for and consumption of refined sugar and the products made with it continued unabated throughout the rest of the century.

The Making of Sweetness

Because sugar went through radical physical and geographical transformations before it entered the marketplace, consumers endowed it with an aura of exoticism and mystery, treating it as they did chocolate: it came from far away, was processed into a form that did not resemble its natural state, had physiologically stimulating effects, and was produced through the labor of Africans and other unfamiliar people. And, because it was initially only accessible to the upper strata of society, sugar signified class and status. All of these psychological factors encouraged the physical desire for sugar and sweets, goods that quickly became affordable to all but the poorest of the poor in sugar-rich England. It took less than a century from the time England colonized Barbados for its citizens to become thoroughly familiar with sugar. Indeed, exploitation of Barbadian land and labor, plus the establishment of domestic refineries in the mid 1600s, enabled Britain to supersede Spain and Holland as a world producer of sugar and helped it reign as a world power.[26]

By 1689, there was a successfully functioning sugar refinery in New York City. By 1800, seven American cities had refineries, including New York, Boston, and Philadelphia (most of the sugar consumed in America before 1789 is believed to have come from the Philadelphia refineries). Three decades later, there were thirty-eight refineries in U.S. port cities (sugarcane was too perishable for overland transportation), including New Orleans, Baltimore, Salem, and Providence.[27]

Satiating consumers' appetites for refined sugar was not easy. Sugarcane proved a difficult crop not only to maintain but also to establish. Propagation began by planting pieces of cane. Much as with potatoes, new growth sprouted from the eyes. The establishment of a new field required the maturation of a series of "stands," usually planted over three years, cuttings from established fields being used to plant successively larger areas. The cutting, tilling, and planting was all done by hand—backbreaking work—until the introduction of mechanized plowing and irrigation systems. In addition, the crops needed time to develop. In Louisiana, the growing season usually lasted about nine months, and the cane had to be harvested from October to January. In the tropics, harvesting 12- to 18-month-old cane lasted from January until June or July, and this older cane yielded more sucrose. Slaves chopped down mature stalks using machetes and hauled them off to be processed.

Extracting the sucrose and turning it into a transportable and edible commodity involved an entirely different set of procedures, which involved a similar amount of labor and, until the advent of steam-powered machinery at the end of the 1830s, defied the implementation of large-scale mechanization. While sugar was often not refined right on the plantation, mills on the premises extracted juice, which could be shipped to refineries along the Atlantic coast of North America and in Europe for purification, granulation, packaging, and transportation to domestic and foreign markets. The key steps in processing cane sugar included extraction, crystallization, and refining. Extraction separated the usable sucrose from other parts of the plant, producing a juice that was then boiled to a thick syrup. After cooling, the syrup was boiled again until it crystallized, or "grained." The purer the sucrose extracted, the more readily the substance could be crystallized (granulated) and refined (purified). Refining continued by separating purer sugar from molasses after it had been clarified using ox blood, lime, or sulfur. After 1855, this was accomplished using the vacuum steam pan, an innovation that enabled sugar's crystallization at a lower temperature, and centrifugal force (perfected around 1860), which could spin out remaining impurities. Before these improvements, refiners practiced the "open kettle" method of refining, which involved using four kettles that boiled sugar in successively purer batches—a crude process that produced brown sugar, explaining why people prized sugar that approached whiteness and purity, why recipe books explained how to refine raw sugar oneself, and why the purest sugars commanded such high prices.[28]

In the early 1830s, mechanization was applied to some of these processes, but typical refineries continued to rely on manual labor for sugar refining. Kettles or copper boilers held a mixture of sugar and bullock's blood or egg white, which was boiled for three and a half hours. A "panman" oversaw this clarifying process, skimmed off the rising impurities, and determined when the syrup was to enter the next stage. He then filtered the syrup through blankets, poured it back into clean pans, and boiled it again to a temperature hot enough to alter it chemically and allow for its crystallization, which occurred after it was cooled and stirred. Finally, a worker packed the refined sugar into molds and baked it. The result was called "loaf" sugar.[29] Refining required both the discernment of skilled and also the muscle of unskilled laborers, and it produced sugars varying in color, flavor, and quality, from, as Sidney Mintz describes, "syrupy liquid to hard crystalline solid, from dark brown ('red') in color to bone white," "and in degree of purity from slight to nearly 100 percent."[30] Nineteenth-century mechanization enabled the production of whiter and purer sugars. Although variations in sugar still existed (from muscovado to brown to confectioners' triple-refined white), steam-driven refining processes delivered a more uniform product and the whiter, cleaner-looking, and hence more desirable form of sugar people craved, in contrast to what the majority had formerly been able to afford—thick, dark, syrupy molasses, literally, the dregs of sugar.

Domestic Sugar Sources

Americans initially enjoyed little success in developing sugar sources of their own that would make them less reliant on imports from Britain and other countries. Domestic sugar production was one possibility, but attempts to grow sugar in North America in the early eighteenth century failed. What looked like a promising business begun in Louisiana in 1756 was soon thwarted by severe weather, the French and Indian War, and the ceding of Louisiana to Spain by the French. The Louisiana sugar industry finally produced modest supplies of sugar by the late 1790s, but its production was sufficient only for local consumption. Sugar cultivation expanded during the first decade of the nineteenth century after the Louisiana Purchase in 1803, when there was an influx of new settlers and cheap labor—mainly Haitian immigrants who had previous experience with sugarcane. Even though Louisiana sugar made the United States slightly more self-sufficient, however, it fell far short of supplying all America's sugar needs. Moreover, sugar produced in European colonies still cost less and continued to be shipped to America.[31] Not until the development of a viable alternative to cane, beet sugar, the annexation of Hawaii, and the acquisition of the Philippines and Puerto Rico after the Spanish-American War—all of which occurred in the late nineteenth century—was America able to extricate itself from its dependency on foreign sugar sources. However, by the end of the nineteenth century, imported sugar had become so much cheaper that America continued to get the bulk of its sugar supplies from overseas.

Substitutes for refined sugar produced from sugarcane had their drawbacks. Maple sugar, for example, was a perennial alternative, but even though sugar maples were indigenous to North America, they only grew in certain areas and yielded limited supplies of sugar. Also, like honey and molasses, maple sugar was a less versatile material than crystallized cane sugar—because of its moistness, it could not be shaped into different kinds of candies, and because of its distinctive flavor, it could not function as an anonymous sweetener in tea or coffee. People did substitute molasses for refined sugar, but primarily in the South, where it was an established component of culinary culture.

That entrepreneurs had invested in maple sugar production with disastrous results did not deter its promotion.[32] A writer in the *New York Magazine* in 1792 argued that the sugar maple was the solution to the evil of slavery on British sugar plantations. Americans should take advantage of the bounty of the "beneficent" trees, especially since sugarcane was fraught with so many problems, from its propagation to its politics: "It is a tender plant; it has many enemies, and requires constant care and labour to defend it from numerous accidents: add to these, the painful efforts that the preparation and manufacture costs to the wretched Africans."[33]

For the most part, however, American investors and consumers alike ignored such pleas, much as the British had paid little attention to their more outspoken abolitionists. The United States "was rather able than likely ever to manufacture [maple sugar] in a considerable quantity," an 1810 census of U.S. manufacturing conceded, even though it had "been so well refined as to have been served to the largest circles of foreign and American evening visitors at the house of the late president Washington."[34] Cane sugar did not need such recommendations. It represented new fashions, new technologies, new wealth, and new possibilities, whereas honey and maple sugar symbolized the past.

Yet scientists, agriculturists, and capitalists continued to experiment with other more acceptable substitutes for refined sugar. In truth, the promise of great profits was the greatest motivator: clearly, the market for sucrose both in America and abroad was rising unabated, limited merely by its affordability and availability. While some suggested substitutes for cane sugar, few people advocated going without sweeteners all together. Given the "very extravagant price of sugar," a writer in *New York Magazine* in 1795 suggested using purified treacle "for the purpose of obtaining a succedaneum equally pleasant and salutary, and capable of being produced at a moderate expense," but admitted that his concoction was not "equal to the best sugar."[35]

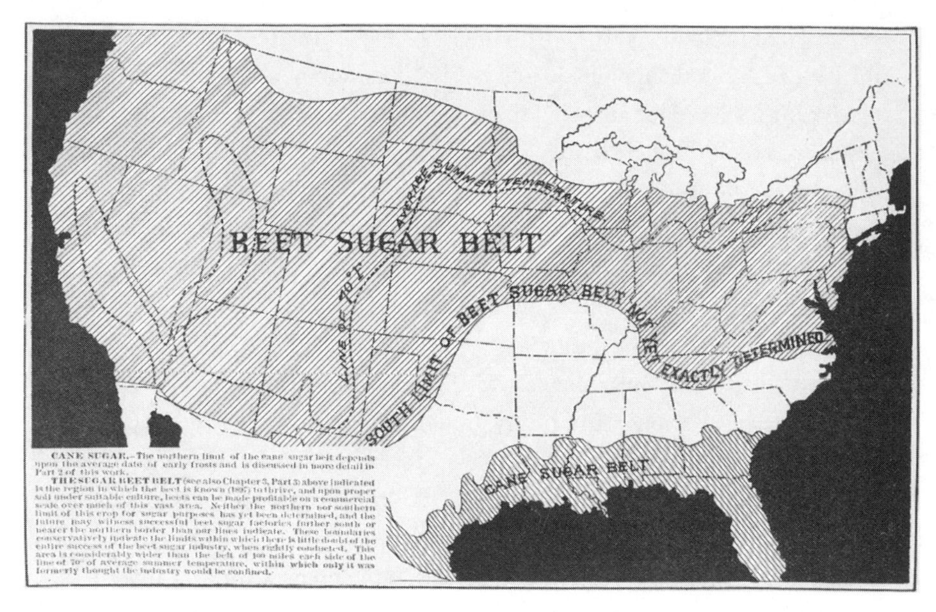

FIGURE 5. The sugar beet, a much hardier plant than sugarcane, could be grown in many more areas of the United States. Herbert Myrick, *The American Sugar Industry* (New York: Orange, Judd, 1899), 72. Library Company of Philadelphia.

Sugar from sorghum, a grain similar to Indian corn that was introduced into the United States around 1854, was tried later, but it, too, proved an inferior replacement, and people turned to it only in dire circumstances, such as during the Civil War, when it served as a substitute for molasses.[36] Experiments with other sugar sources ranged from the logical to the bizarre. There were enough sugar maples in America to inspire continued attempts to streamline production throughout the eighteenth and nineteenth centuries. Others tried with substances derived from corn stalks, sweet potatoes, pine trees, acorns, manna trees, chestnut trees, milk, and even licorice root.[37] In the end, however, only the sugar beet offered a viable alternative source of refined sugar.

Producing sugar from beets became a successful commercial reality in the mid nineteenth century and quickly enabled and accommodated skyrocketing consumption rates of sugar and confectionery. While the plants required the same kind of processing as sugarcane, beets could grow in temperate climates, in places too cool for sugarcane. In the United States, this meant a 2,000-mile-wide swath across the entire country—a much larger area than the sugarcane-growing zone of Louisiana and a few isolated areas in Texas, Georgia, and Florida (fig. 5). By providing a much more affordable and prolific alternative to cane, beet sugar production revolutionized not only the sugar and related industries but the American diet as well.

Andreas Marggraf, a German chemist, discovered beet sugar in 1747, and France started using it in 1806.[38] By 1836, the Beet Sugar Society of Philadelphia had been founded to promote the production of beet sugar in the United States as well—a tempting endeavor, since a contemporary account, Edward Church's *Notice on the Beet Sugar,* claimed that France was producing from 60 to 80 million pounds of beet sugar a year, enough to make that country self-sufficient. In 1837, the second American edition of Church's work was published simultaneously in many American cities, spreading the word to places like Philadelphia, New York, Buffalo, and Boston.[39] Production figures were impressive: already by 1858, 16.6 percent of the total world sugar production came from beets. By 1873, this figure more than doubled, to 34.1 percent, and by 1893, beet sugar accounted for more than half of world sugar production, 54.3 percent.[40]

Until the late 1870s, the United States contributed little to these figures, mostly because it already had a regular and sufficient supply of raw sugar from the West Indies, and because, until then, most government incentives focused on improving sorghum sugar production. In 1838, David L. Child tried to extract sugar from beets in Northampton, Massachusetts, but had little success—the beets he grew yielded only 6 percent sugar. Subsequent attempts occurred decades later, beginning in 1863 and continuing until the 1870s, in places such as Illinois and California.[41] By the late 1880s, American producers finally gained a foothold in the beet sugar market. In 1896 alone, however, cane plantations in Hawaii produced four times the yield of either Louisiana cane sugar or California beet sugar. All in all, by 1906, fifty-three U.S. plants were producing over 300,000 tons of sugar annually. The U.S. government's support was manifested by the Dingley tariff of 1897, which both lessened reliance on imports and increased U.S. manufacturers' profits by protecting domestic cane and beet sugar, and the country eventually became a key world producer.[42] At last, American sugar producers could supply the demands of American consumers.

CHAPTER TWO

Sweet Youth

∴ *Children and Candy* ∴

On 17 August 1730, Nicholas Bayard opened one of the first establishments in North America that both processed sugar and sold goods made from it. His advertisement in the *New York Gazette* informed readers that his "Refining House" also offered four kinds of refined sugar and "Sugar-Candy."[1] Consumers could certainly get sugar in early eighteenth-century America, but they bought it from a dry goods store, a grocer, or a spice merchant. It came in hard, compacted twelve-inch-high cones. Tongs were used to break off chunks, which a special grater then reduced to granular sugar. People known specifically as confectioners who sold confections did not appear until later in the century.

Sugar often figured in early confectioners' wares as a subsidiary ingredient; people came to these shops in search of exotic imported fruit, for which sugar was often used as a preservative. For example, in 1763, thirty-three years after Nicholas Bayard's New York notice, James Marsh, a Philadelphia grocer, offered a few confections to his patrons, who were an elite clientele. His advertisement promised customers typical eighteenth-century confections, including raisins, currants, "pruans," figs, almonds, citron, green sweetmeats, chocolate, three kinds of sugar, and cordials.[2] In 1775, a Philadelphia confectioner,

Patrick Wright, advertised an even more assorted array of treats, including two kinds of comfits, sugar plums, three kinds of sugar candy, fruits, jellies, marmalades, macaroons, and biscuits.[3] The 1811 estate inventory of Augustus Lannier, a New York City confectioner, listed two hundred bottles of "sundry cordials," seventy-six bottles of "sugar works" and "nonpareils," eleven boxes of "Sugar Things," and "Sundry Potts of Sweetmeats or Preserves." This was in addition to paper novelties, flavorings, nuts, and his equipment, which included copper kettles, a distilling apparatus, a marble table, a mortar and pestle, and many other basic items necessary to the confectioner's trade.[4] The market for and supplies of such treats was so limited that even by 1816, only twenty confectioners operated in the entire city of Philadelphia.[5]

One of these early and exclusive Philadelphia confectioners was Sebastian Henrion, whose business was representative. Although Henrion also sold imported chocolate and bonbons, he mostly purveyed nuts, syrups, and fresh or candied fruits in his early business—all very precious and desirable goods. In April 1826, for example, Henrion sold over $8 worth of goods to a customer, including four pounds of almonds at more than thirty-one cents a pound, four pounds of raisins at the same price, twelve oranges for $1.25, and four pounds of figs for 75 cents. A few months later, the same man bought another two pounds of almonds and two pounds of raisins.[6] On these two visits to Henrion's shop, he spent a total of $10.06, leaving him with an outstanding balance due of $187.50—an extravagant amount of money at the time, more than an average laborer's yearly wage.[7]

Penny Candies

The nature of the confectioner's enterprise and the high class of his customers remained consistent until the late 1830s, when technological advances and the accessibility of sugar expanded the market. After that, the confectioner's business began its transformation into a completely different institution, favoring not the palates of the rich but the pleasures of the working class, no longer suiting only the tastes of adults but those of children too. As a result, rather than being a site for the display of grown-up prestige, the local confectioner's became a venue for the children of early American capitalism. Of course, fine confectioners remained, but the candy store became an enduring institution by catering to a very different set of customers and their needs.

The success of penny candy epitomized the transformation of the confectioner's shop into the candy store. And the candy store helped acclimatize the lower and middle classes to the idea of confectionery consumption. Cheap, brightly colored, fanciful pieces of hard-boiled sugar, penny candies enchanted children, who saw them as dazzling gemstones in the glass jars of the corner store. What is more, these sweets were the first material goods that nineteenth-century children spent their own money on, making them so important that candy store owners later in the century relied almost exclusively on the patronage of children to keep them in business. These kinds of confectioners made candy accessible to children with the help of machines that made sweets more rapidly and cheaply. In the hands and mouths of the elite, fancy sweets—fine French bonbons, pastilles, and nougats—served as status symbols, expensive trifles. But the penny candy, more accessible to larger groups of younger people, represented freedom and pleasure to children, and the danger of independence to reformers who wanted to control people's intake of sugar, especially in the form of confectionery. The success of penny candy by the end of the century took refined sugar out of the mouths of the elite and counted among the ways in which sugar, sweetness, and candies became associated with the weak rather than the powerful.

Hard candies of the penny variety did not appear in the marketplace until well into the nineteenth century. When Nicholas Bayard advertised in 1730, the "sugar-candy" to which he referred was most likely medicinal—boiled with water to a high temperature, mixed with herbal extracts, and then solidified to a brittle hardness when cooled. These proto-candy offerings came from a long-standing tradition of prescribing sugar syrups and sugared tablets for their palliative effects. *A Queen's Delight* (1662) contains very early recipes for making something like the hard candy we know today. For example, "To make Lozenges of red Roses," was very simple: "Boil your Sugar to Sugar again [sugar syrup to the hard crack stage], then put in your Red Roses being finely beaten, and made moist with the juyce of a Lemmon, let it boil after the Roses are in, but pour it upon a Pye-plate, and cut it into what form you please."[8]

Hard penny candies, which did not become ubiquitous until the mid 1840s, more than 100 years after Bayard's advertisement, directly descended from medicated lozenges, pills masking often bitter medicine in a palatable hard sugar coating. Counting traditional uses, nineteenth-century American cookery books also recommended certain sweetened foods for the sick and invalid, reinforcing yet again confectionery's growing association with society's weak. "Treat" as a noun and a verb had become conflated in one product.

In 1847, Oliver Chase, a Boston druggist, developed a machine resembling a contemporary printing press, consisting of a flat bed and two rollers embossed with pill shapes that enabled him to manufacture his medicated drops more efficiently. By working a concoction of gum arabic and brown sugar to just the right consistency, Chase was able to run this sticky "dough" through his press, yielding "a stream of lozenges, each perfectly shaped, tumbling down onto his work table in rapid succession."[9] Known as "cold sweets," such lozenges retained their chalky texture and association with medicinal remedies, much the way peppermints do today.

Inventions like Chase's device and J. W. Pepper's lozenge-cutting machine, patented in 1850, made use of refined sugar's capacity to harden into a molded shape after being boiled. At the same time, distillers provided apothecaries, perfumers, and confectioners with a variety of essences and extracts to enhance their products. With these improvements, refined sugar made its way in concentrated and easily edible doses to many new consumers. Confectioners seized these opportunities, omitting the medicines, adding more refined sugar, and beginning mass production of hard candies aimed at pleasure rather than pain.

Reformers, however, expressed concerns about confectionery in an attempt to make people circumspect about this newly accessible substance. Even before hard candy became widely accessible in America, commentators put out warnings echoing those already familiar in England. An article in *The Friend* of 1834 claimed that the ingestion of confectionery by poor children would lead to "intemperance, gluttony, and debauchery."[10] The author believed that unlike the rich and refined, the poor did not have full control over their actions or destinies: "Poverty, in some instances, certainly leads to vice," and candy counted among the things that would tempt them.[11] *Parley's Magazine,* a frequent publisher of articles extolling temperance, warned its young readership in 1835 to avoid confectionery "as if it were poison," and continued, "No boys and girls are more healthy than those who live in places where they can get no confectionary."[12] In 1837, *The Colored American* referred to candy shops as "hot beds of disease," "filled with putrid rottenness," and implored parents to cease giving their children money to spend there.[13]

Temperance supporters patterned their critiques of sugar on their critiques of alcohol, figuring that sugar could be just as addictive as liquor and might gradually engender bad habits. (After all, people made rum with it, too.) In fact, some even believed that a childhood sweet tooth manifested itself as alcoholism in adulthood. For example, the *Temperance Advocate and Cold Water Magazine* featured a story in 1843 about a boy named Henry Haycroft who lived in a town where "almost everybody drank, from the minister in the pulpit, to the little boy who eat[s] the sugar out of the bottom of his father's toddy-glass." The story describes how Henry, seduced by his father's "range of decanters containing liquids of varying hues," took his first drink as a youth. As a teen, Henry took his first real drinks, buying nips of peppermint cordial for a cent apiece and imbibing them with a friend. "The liquor tasted so sweet, that they all took a large quantity."[14] The author underscored his analogy of alcohol and sugar—both peppermint-flavored, sweet, in glass containers in various hues, and available for a cent—by following the young duo stumbling into public and buying candy at a candy stand. A cautionary tale of nascent intemperance, the story expressed key concerns of reformers, concerns which recurred throughout the century.

Of Men and Machines—The Producers

Most people did not heed the warnings of reformers, and confectioners continued to improve their machinery and marketing, aware of burgeoning consumer desire and the potential for profit. By the mid 1860s, confections in some form were available and accessible to nearly everyone. In 1864, the periodical *Once a Week* credited steam power with transforming the confectionery trade, observing, "what was once an article of luxury for the use of the rich, has now become an article of necessity almost for the children of the very poor."[15] Inventions such as the revolving steam pan, Bessemer's cane press, the vacuum pan, and other devices using steam power democratized both refined sugar and the confectionery made from it.[16] By the 1880s, even those who lived in remote towns could sample the goods of the local candy man.

Steam power initially enabled tool companies to manufacture equipment for the small confectioner. Only later, in the 1870s and 1880s, did entrepreneurs apply steam power to large-scale candy making. For a relatively small expenditure on a basic press and one or two interchangeable shaping rollers, a man could acquire the equipment necessary to start a small but potentially profit-

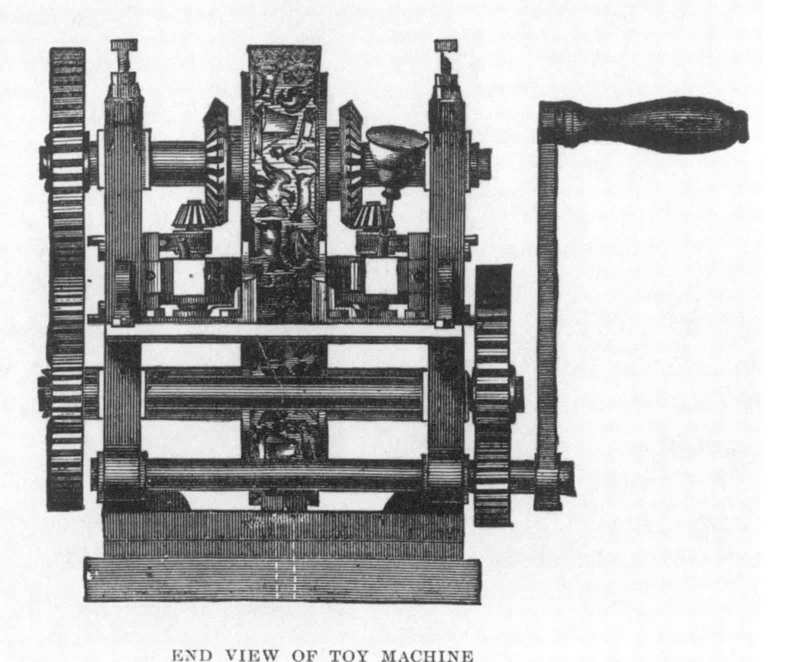

FIGURE 6. Heavy-duty candy-making equipment enabled confectioners to increase the amount and variety of their output. These "toy" machines, modeled on the printing press, had interchangeable rollers that literally cranked out hundreds of different candy shapes. "Mills' Excelsior Clear Lemon Toy Machine," Chapman & Smith Co., *Catalogue* (Chicago, 1899), 142. Library of Congress.

able candy business, especially if he ran it in conjunction with a bakery or tobacco shop. Once he determined the proportions of sugar, water, flavoring essences, coloring agents, and boiling times (and much of this came through experience, as the nature of sugar boiling changed with outside temperatures and humidity levels), a candy man could literally crank out all varieties of shaped, colored, and flavored candies. Thomas Mills & Co., a Philadelphia confectioners' equipment manufacturer founded in 1864, supplied many confectioners with presses, called "Toy Machines." A basic "Clear or Lemon Toy Machine" (fig. 6), which came with two sets of rollers, sold for $150 in 1866.[17]

Replacement rollers could be had for $14 or $16 a set, depending on the intricacy of the castings, the size of the press, and whether or not the rollers carried personalized company names, as did "motto rollers." Independent operators and larger companies alike used this machinery because the technology required to make penny candies remained the same, no matter the quantity produced.

By 1871, the Mills company was directly shipping its Toy Machines all over the United States, to George Miller in Philadelphia, to M. E. Page & Co. in Chicago, and J. Ratta in Galveston, Texas, to name a few. In addition, Mills hired agents in major cities to promote and distribute its equipment. This operation fostered repeat business, because successful candy makers could gradually build up their equipment as they garnered profits. For example, George Miller and Son bought $586 worth of equipment between April of 1864 and July of 1870, usually buying one or a few rollers, candy cutters, or composition molds at a time.[18]

The Mills company counted as one of the few nineteenth-century firms that specialized in the manufacture of confectionery equipment, competing only with similarly large companies in New York and Boston. Because they had sound distribution networks, these firms supplied large and small confectioners all over the United States, enjoying great success delivering the tools needed to make the candy that sated the American sweet tooth. In 1868, for example, Mills's business was so successful that he declined to give a discount to Henry Maillard, a well-known New York City confectioner. Of his Toy Machines, Thomas Mills instructed an agent to "not sell them for less than our Circular price. As we can get the Cash for them here without selling them at a sacrifice." Mills gained his reputation largely outside New York City, where confectioners insisted that he "prove the advantages and good qualities" of his machines against local competitors."[19] Over a 15-year period, his company, using jobbers, supplied distant candy men operating in places such as Oil City, Pennsylvania, Salt Lake City, Utah, Helena, Montana, and Glasgow, Scotland.[20]

Using equipment supplied by Mills and others, enterprising men made cheap candies on their own and started candy businesses requiring little investment capital and no additional employees. Most candy makers, however, hired someone to oversee the vats of boiling sugar, pour the molten mass onto a marble slab, work it with wooden paddles until it cooled to the proper consistency, and transfer it to the press bed of the candy machine, where it was cranked through shaped rollers—much like wet clothes through a wringer—and separated into candies at the other end. While some businessmen used the penny candy trade to attract patronage to other parts of their stores—the newspaper rack, the soda fountain, the tobacco counter—others relied solely on the sale of candy for their livelihoods. Because penny candies could be made and sold in bulk, but by nature yielded a very low profit margin, those working exclusively in the candy business had to increase that margin in any way possible. Some used poisonous dyes and artificial flavors to make low-grade candies, which reformers did not ignore.

⁓ *Sweet Urchins—Children* ⁓

By the 1860s, children buying penny candies, rather than fashionable people seeking preserved fruits and comfits, made up the confectioner's customer base. Shiny bright yellow, orange, green, and red candies packed in jars perched on window sills and filling bins under glass countertops attracted young consumers. Glass windows, display cases, and jars that enhanced the eye appeal of penny candies were increasingly affordable and trained new generations of shoppers to both appreciate the vast array of material choices made possible by mass production, and also to learn how to discern the relative merits of these various goods by sight. "Within a generation people have learned more of goods and products, and of uses and comparative advantages by seeing and comparing displays, than they ever knew. Our shop windows are a kindergarten for grown-up people, as well as little ones," a trade journal observed in 1900.[21] People learned in this "kindergarten," this basic school of commerce, that the abundance of goods before them could be had for the right price. William Leach has noted that display glass, being a visual medium, "democratized desire even as it democratized access to goods."[22] This equal-opportunity gazing discriminated neither by class nor age. Shopwindows encouraged looking at the commodified world, and the displays behind them were merely three-dimensional versions of the vibrant print ads in magazines. Within the shops themselves, counter urns, exposition jars, globe jars, and section jars numbered among the many glass containers positioned atop a confectioner's counter, full of jewel-like candies (fig. 7). While adults experienced this commercial effect largely through the pane of the department store window, children learned it within the enclosed milieu of the candy store (fig. 8).

Children invested both time and money buying candies, enacting consumption rituals that prepared them for their lives as adult consumers. Some spent hours, day after day, in front of a candy counter, surveying all the choices at hand. "Perhaps the most difficult decisions were those made in front of candy counters," one man reminisced. "You would feast your eyes from left to right, from right to left, from front to back, and even diagonally."[23] In these milieus, even very young children began to understand the process and pleasure of consumption, a way of understanding the world that would remain with them as they became adults with even more purchasing power. Children also came to realize what it meant to want, to yearn, to be tempted by treats simultaneously accessible and inaccessible to them. They could look at these candies all they wanted, and they were encouraged to do so, as candy makers posed the treats enticingly behind store windows within sparkling glass jars, "a mingling of refusal and desire that must have greatly intensified desire."[24] But children could eat them only if they had enough money (fig. 9). "There was a tacit understanding between proprietor and customers: children without money to spend had to stay outside, but those with only a penny were welcome to take all day if they wanted, picking out their treat, savoring every morsel, and hanging around afterward," David Nasaw notes.[25] This "tacit understanding" inculcated in children the principles of capitalist pursuits (democratized did not necessarily mean democratic), teaching them at an early age that entrée into the world of consumerism came at a price—an important lesson for a mere penny.

Children had many ways of getting money to spend. Fortunately for candy shop owners, by the 1870s—when sugar had become most plentiful, equipment affordable, and local confectioners familiar—more children worked for wages than ever before.[26] Some peddled goods in the streets, like newspapers, flowers, and even their own candy. They brought their earnings home, received back enough to buy the next day's supplies, and sometimes retained an additional amount of spending money if their parents could spare it.[27] "Since parents resisted making allowances entitlements, preferring to dole out small sums for treats, to have spending money that they controlled most children had to earn it."[28] Other children worked not because their families needed the

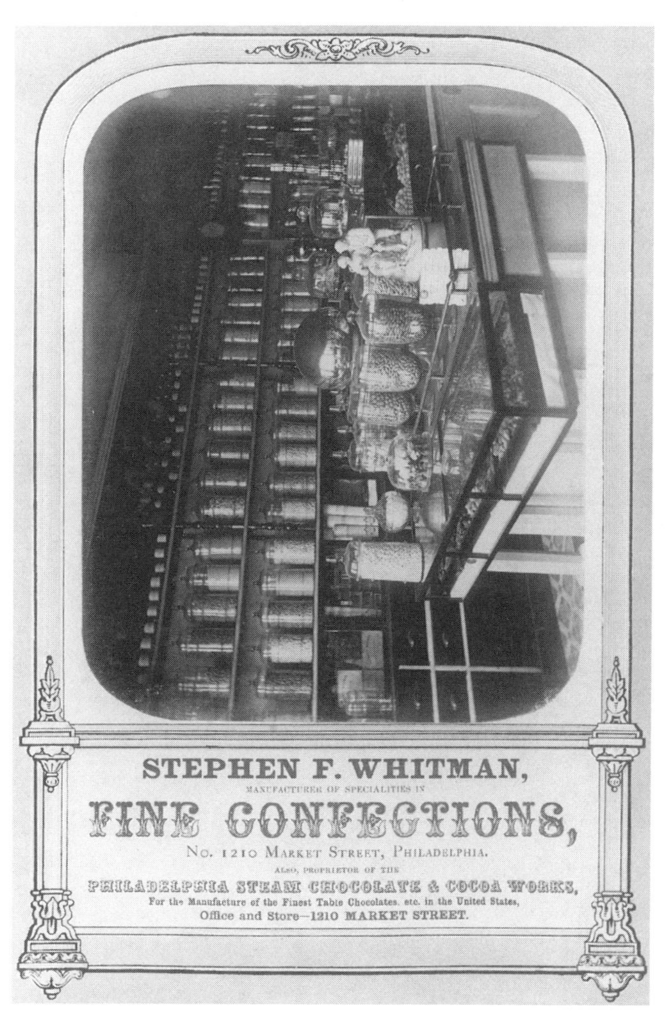

FIGURE 7. Candy stores were treasure chests for children and offered innumerable treats from which to choose. Interior of Stephen F. Whitman's store, Philadelphia. *Gallery of Arts and Manufactures of Philadelphia* (Philadelphia: Wenderoth, Taylor, & Brown, ca. 1870). Library Company of Philadelphia.

FIGURE 8. Candy store window displays often captured the attention of children and lured them inside. G. W. Anderson Confectionery, Albany, New York. Trade card, ca. 1890. Warshaw Collection of Business Americana. Archives Center, National Museum of American History, Smithsonian Institution.

FIGURE 9. With the help of penny candies, children learned the difference between the "haves" and the "have-nots" at an early age. "The Candy Shop," *Arthur's Home Magazine* 22 (1868): 255. Library Company of Philadelphia.

money but in order to gain a degree of independence or to purchase things for themselves that their parents would not buy.[29] Even the poorest children in the cities, whose families desperately needed money, obtained the errant penny. They begged. They ran errands. They earned bounties by collecting vermin. They even searched the floors of trolley cars.

Significantly, penny candies counted among the first things that American children ever spent their own money on, since merchants offered these treats in their shops long before other marketplace diversions appeared. But penny candies remained attractive to children even decades later, at the turn of the century, when other cheap entertainments like arcades and movie theaters were available, especially in the urban landscape. Candy shops became places where youngsters gathered and socialized among themselves, away from parental control, where "they were allowed, even encouraged, to act more grown-up than was good for them."[30] The pragmatic retailer understood that his financial success relied on the patronage of children; he was at their mercy. "A Curious Candy Store Boycott" took place in a rural Pennsylvania town at the turn of the century, for example, because the sole candy shop proprietor forbade loitering. "The children are organized, and have held several torchlight processions," the *Confectioners' and Bakers' Gazette* noted uneasily.[31]

But the power dynamic worked both ways, and confectioners implemented their own strategies for "attracting their trade." While the skillful use of public display space most immediately attracted children, confectioners also devised other, more covert tactics. One trade journal suggested giving change in pennies, which people would deem more expendable: "Shiny pennies are the children's delight, and, if they do not get them immediately, rest assured that within a day or two they will be teased away from the older people and if you have secured the confidence of the little ones you will get every one back again." This author estimated that if a merchant befriended fifty children in a year, who "turned in 500 pennies each, which is not a large number for children of parents in average circumstances," his business success could be "traced directly to your efforts to please the children."[32] While confectioners could also rely in part on occasional trade from these children's parents, they mostly depended on the estimated $5 a year each child spent on candy.

Especially for lower-class children, penny candies functioned as an important cultural staple, offering infinite fantasy worlds in addition to something delectable to eat. They made possible the seemingly magical exchange of one thing—a penny—for an abundance of goods—a bag of candies. Furthermore, the penny enabled browsing, which often lasted a lot longer than the treats themselves. One particularly detailed reminiscence describes this pleasure of choice, and merits quotation at length:

> There was the old candy-woman, smiling enticingly over her counter (a cross candy-vendor will never have many customers among the little folks), and there was the group of children considering the possibilities

> of a penny. What could they get the most of for a cent. Would a sourball or a mintstick last the longest. How many peanuts did she give for a penny, they asked of the old lady, and how many gum-drops for the same sum, and how big were her penny-cakes of maple-sugar. All these considerations, and I don't know how many more, were duly weighed over this penny as they had been dozens of times before, I doubt not. What investment was at last decided upon I did not wait to see, but thought, as I turned away from the scene, how much more value and how much more pleasure there was in the penny of childhood than in the millions of maturer years.[33]

Children practiced a great deal of discrimination when buying candy. In the process, they learned how to be competent, educated consumers who would grow up to be the enthusiastic customers sifting through the many mass-produced goods in department stores and public markets.

That people often referred to and thought of these candies as "toys" further reinforced their solid position in the children's market. Marketing strategies often blurred the line between "real" toys and candy toys, associating them all with pleasure and entertainment. For example, the display for Smith's Phen-Hunch Marble Gum resembled current games of the time (fig. 10). The product, simultaneously a toy and candy—marbles and gum—could be played with and then eaten. Whether children actually played with most of these candies like toys is unclear, but the candies certainly mimicked shapes of other common toys, including miniature horses, guns, dolls, dogs, flowers, and stars, and so imitated things already familiar in their material universe. Also, because these toys readily disappeared into the mouths and down the gullets of their little consumers, they ensured years of customer loyalty from the child with a mature sweet tooth, who would return for more. For example, the most reliable customer of Hepzibah's cent shop in Nathaniel Hawthorne's *House of the Seven Gables* is Ned Higgins, the "little devourer of Jim Crow and the elephant," who developed a penchant for the gingercakes after getting his first ones free.[34] Even though children cut down on their candy consumption when they reached their teens, they could be counted on as avid consumers up to at least the age of ten.[35]

In many ways, the candy toys, like other toys of the time, represented children themselves—merely miniaturized versions of their parents. Children en-

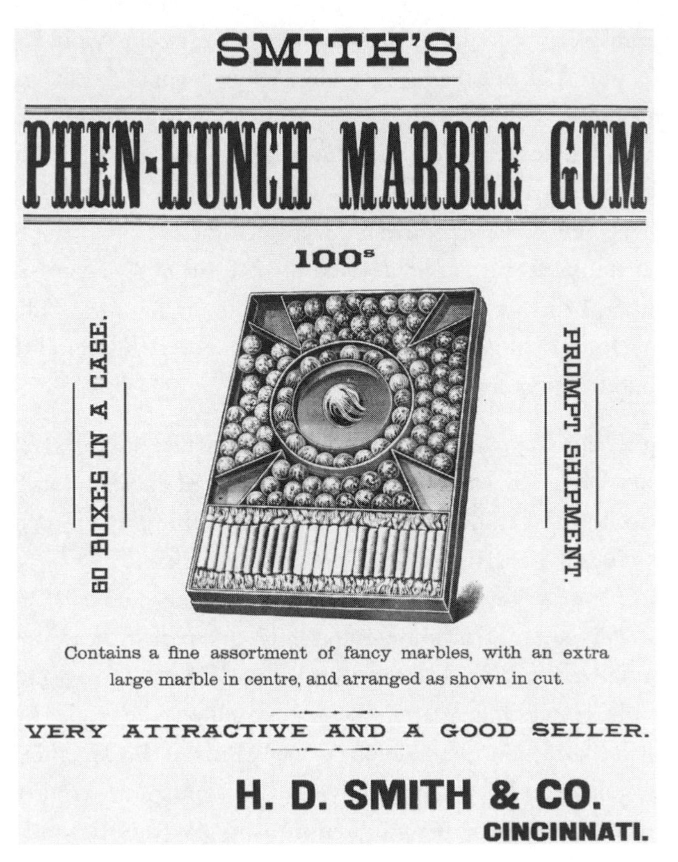

FIGURE 10. Making candy novelties that looked like games was one strategy confectioners used to attract the trade of children. Phen-Hunch Marble Gum, *Confectioners' Journal,* February 1890, 34. Library of Congress.

joyed small, ephemeral versions of adult commodities and through selection and purchase participated in adult, consumer-oriented behavior. "Poring over the selections, choosing what to buy and where and how much, children transcended their minute size and inferior status to assume quasi-adult dimensions," David Nasaw notes.[36] The act of purchase may have provided more enjoyment than eating the goods themselves, and it certainly let children fantasize about adult life. But the commodity forms themselves remained equally important to children. After all, they did not spend their precious pennies indiscriminately but made deliberate choices as sophisticated consumers. Candy shop owners employed a number of strategies to encourage children's trade, because they could just as easily spend their pennies on pieces of lace or cheap cap guns.

Sophisticated strategies targeted specific classes of children. A candy man strived to forge quasi-personal relationships with his young customers. In order to cultivate "friendships," merchants offered a small toy with each purchase.[37] Such tokens transcended their purpose as novel "jimcracks" and acted as the medium through which businessmen, by giving "gifts," created mutual obligations on the part of individual customers.[38] The little toy proved an efficacious marketing tool because, in the words of one contemporary advisor, like other advertising novelties, it appealed "not to reason, but to the heart, to the emotions, to sentiment, to good will on the basis of implied acquaintanceship between advertiser and potential customer."[39] Ironically, of course, the premium—the gift object that established and concretized these individualized "acquaintanceships"—was itself not unique but one among thousands of identical copies, all the output of mass production. For example, W. C. Smith, of Buffalo, New York, manufactured and sold "confectioners' toys for 'penny goods,'" by the gross, including "Puzzle Whistles, French Puzzles, Pop Guns, Tin Dishes, Stamped Spoons," and many other items appealing to both girls and boys.

Although they were often thumbnail-sized and made of cheap metal, these toys gave children tangible incentives to buy candy, because the idea of getting them "free" whetted juvenile appetites,[40] something not lost on twentieth-century marketers of Cracker Jack and sugarcoated cereals. These premiums helped children begin to appreciate the power of possessions, even if they were too young to grasp the disparity between the premiums and finely crafted goods. Like adults, children of all classes busied themselves by amassing a collection of paraphernalia that provided evidence of one's life experiences and economic status.

Candy premiums promised a (literally) scaled-down version of a materially rich life to those who would never be able to afford it. Trade journals like the *International Confectioner* showed the array of toys a confectioner could offer with his candy. However tiny, tokens existed as entire worlds by themselves, worlds representing luxury in minuscule detail. Dowst Brothers Co. of Chicago (later manufacturers of the enduring Tootsietoy line of metal cars) advertised "Metal Novelties for Penny Prize Goods." Dowst described its rabbit, for example, as "ready to spring. A plump, alert figure, with fine head and long, erect ears. Feet, tail and fur all well defined. Three-quarters of an inch high and nine-sixteenths of an inch wide. Finished on both sides. One of our most artistic productions. Finished in gilt or silver or in colored enamel." Another novelty, the Ladies' Shoe, with the "Latest French heel," came in six different finishes and appealed to those with nascent discriminating tastes. (Thorstein Veblen identified such shoes, with "the so-called French heel," as a sign of the "enforced leisure" particularly enjoyed by the leisure class.)[41] A final example, the American Yacht Cup novelty, most blatantly embodied the material world these young consumers would never know: "This is an exact reproduction of the famous Lipton Cup. It has an open work handle, and is a perfect reproduction to the smallest detail. It is an inch and one-eighth in height. Finished in silver or gilt. It will immediately attract attention." That a lower-class child's only interaction with such fine goods occurred through miniature imitations, "perfect reproductions," made the real objects wondrous and unreachable at once. As talismans, they conjured up distant worlds and represented finery forever out of reach.

Toy novelties delayed immediate material want by presenting longed-for commodities in tangible form. Stimulating and then satiating desire put merchants in good stead with their young customers, and these "little souvenirs" could be seen as "each a sales ambassador for opening up better trade relations."[42] They also offered more possibilities for fantasy play than did their larger, more realistic counterparts. One man described the power of the miniature and the pleasure they brought, "because I could play expansive games with them." Little dolls, for example, "had no meaning in themselves, they were not like the large dolls to my sisters, but they made good pawns in imaginative household games, and I could put them into tin railway trucks and in the wooden cities that I snowed over."[43] Premiums did more than encourage one to purchase penny candies. They also enabled even poor children to participate in a commodity culture that could reward and exploit them. The premiums tapped into a desire—for sweet things, for material goods, for fantasy worlds—and allowed children to placate that desire through consumption. Yet the satiation of desire proved as fleeting as the candy itself and brought only more desire and more consumption.[44] The toys, miniatures, were fitting premiums for children whose worlds were "limited in physical scope yet fantastic in [their] content."[45] The toys encouraged an intense desire for possession, while the sweets triggered intense physiological desires for consumption. The two goods worked together, combining intensity with ephemerality, resulting in a cycle that created more longings by their very placation.

Candy's situational packaging—the glass display jars in the confectioner's shop—only intensified this desire. Susan Willis has remarked that in capitalism the packaging is as important as the product, because it heightens anticipation, which she describes as "the commodity's most gratifying characteristic." She continues: "No commodity ever lives up to its buyer's expectations or desires. . . . The shoddy purchase that does not fulfill its advertised promise promotes the pleasurable anticipation of the next (hopefully less shoddy) purchase."[46] Full of glittering cheap candy and metal toys, the local candy shop was outfitted to encourage and heighten consumer desire among children.

Yet children remained critical buyers, and businessmen took seriously their mercurial preferences. They altered candy shapes, colors, flavors, and labels in order to capture the interests of their young customers. As one tradesman put it, using rhetoric familiar in the fashion trade:

> Older people buy confectionery for its quality, but children buy it for its novelty. . . . We are constantly dressing old candies in new garb. . . . The "jawbreaker" has turned into a "butterball," which is yellow and hard and has the enduring qualities of the old fashioned jaw breaker. The white chewing gum, which has been in existence as long as I can remember, was rather sniffed at by the present generation of children until some one had a happy idea to manufacture it a foot long and on the wrapping paper print: "A foot of gum for one cent."[47]

Shapes changed with seasons, holidays, and even technology. Trade journals advised tradesmen to keep candies "timely": "In the spring we got out regulation lollypops, but instead of the ordinary stick there was a rake, hoe or spade with a handle stuck in the candy. . . . Fourth of July calls for candies in the shape of firecrackers and torpedoes."[48] Emphasis on the new, the changing, the dynamic paralleled the pace of rapid industrialization and the need to habituate people's consumption patterns. Revolving around a seasonal or annual desire for "original" styles, colors, and gadgets, they would fuel the commodity-driven economy.

Merchants also recognized class and age disparities, offering an assortment of novelties for older and more affluent children. Significantly, such novelties stressed long-term investment rather than immediate consumption. Long before Cracker Jack's marketing gimmick, the confectionery supply house of John M. Miller and Son advertised twenty-one different "Prize Packages" containing mixtures of candies and surprises, called "prizes," "novelties," and "gifts." According to their 1876 Christmas price list, some packages held cash and jewelry, while others included popcorn, chromolithograph prints, and fine candies. The United States Mint Box, for example, held all cash and fine candies, with a "Gold or Silver Coin in value from 5 cents to $1.00" in every box. The International Prize and Fortune Box was "among the latest novelties out; filled with Imperial Mixtures, a prize, and tells the Fortune of a Lady or Gentleman." Perhaps most alluring, however, was the Great Watch Novelty, which retailed for 35 cents. It promised a "New Silver Coin" in every other box, and a new watch among every hundred boxes. Each box simultaneously embodied temptation, chance, and potential wealth, channeling desires toward specific status-oriented commodity forms, "surpassing all others in VALUE OF PRIZES, QUALITY OF CANDIES and BEAUTY OF DESIGN IN EXTERNAL FEATURE OF PACKAGES, and are the only GENUINE BOXES PACKED, BEING ORIGINAL WITH THIS HOUSE."[49] Children learned how to invest in their future by amassing seemingly valuable goods—many related to money itself—which they could keep long after they had eaten the candy.

Because novelty boxes sold for from 10 to 35 cents, candy men intended them for children who commandeered more money from their parents, were old enough to have higher-paying jobs, did not need to contribute to household expenses, or enjoyed their own allowances. The tangible things that remained after the candy disappeared were much more substantial than premiums from the penny schemes. Penny candy came with paltry tokens in the guise of interesting toys—tiny reproductions of by-products from a lifestyle most working children would never experience. In contrast, "finer" candies, accompanied by smaller versions of genteel trappings—watches, jewelry, gold and silver coins—came packaged as high-class marketplace commodities, evincing "BEAUTY OF DESIGN IN EXTERNAL FEATURE OF PACKAGES," as Miller's text emphasizes. These counted as more substantial "investments," to be saved and displayed, rather than minuscule, throw-away, easily lost trinkets.

Confectioners used other ploys to attract the trade of moneyed children. They based these strategies on principles of cut-throat competition that encouraged children to practice adult roles. For example, one candy man "awarded" a pony to the child who brought the most business into his shop in a two-month period. This marketing scheme targeted and rewarded children's desires to be entrepreneurs, showing "so much hard work and enterprise on the part of the youngsters."[50] Businessmen correctly assumed that the kinds of advertising campaigns pitched to adults worked as effectively with children. The pony strategy was a variation of premium-based marketing directed at women, whom companies rewarded for either inducing their friends to purchase their product or motivating local grocers to stock their brand of food. What the public construed as merchants' goodwill in the form of rewards for patronizing a store really just constituted a marketing technique aimed at a population with a nascent taste for conspicuous consumption in a culture that celebrated stiff competition and hard work.

Other aspects of a child's world also nurtured acquisitiveness and reinforced the lessons learned at the candy shops. Popular literature, such as *The Youth's Companion,* owed a good deal of its success to the premium system. This serial printed entire issues illustrating and describing special bonuses one could earn by selling enough additional subscriptions to friends. Premiums offered in 1892 included an Adjustable Iron Mitre-Box and Saw (worth the enrollment of two subscribers), the Pocket Companion Tool Holder, No. 5 (one new subscriber plus 10 cents), the Domestic Tool Case (one new subscriber plus 15 cents), and The Daisy Printing Press and Outfit (one new subscriber). Remarkably, the magazine also offered the Outfit for Making French Confectionery (attainable for one new subscriber and 10 cents), whose illustration featured all the signs of fine confectionery: a box with a lithographed cover of a girl and lined with a lace doily, a pair of fancy tongs, and different shaped candies both wrapped and unwrapped. "The outfit consists of a Nickel-Plated Dropping Wire, packages of Vegetable Coloring, Compounds, Flavoring Acids, Waxed Paper, 2 lbs. Confectioners' Sugar, and Lessons in French Candy-Making," the ad copy stated.[51] Even though the actual ingredients provided would not really make "French Candy," the rhetoric accompanying the suggestive illustration promoted the desire to have fine confectionery and was aimed at children who could appreciate it as a prestige item. The luxuriousness of genuine French confectionery came from the way experienced hands (not young entrepreneurs) worked with high-quality ingredients (not "compounds" and "flavoring acids"). The Outfit for Making French Confectionery, like the other premiums offered, articulated the romantic notions of bourgeois adults regarding hand craftsmanship that they themselves would never actually have to perform. It is remarkable that advertising taught even the youngest children to appreciate material goods beyond their use value. Objects could help attain and affirm social status, they could engender worlds of romanticized occupations, and, most important, they could help one to acquire more objects.

In addition to prize packages, contests, and premiums, confectioners also used packaging itself to entice children and their parents into making greater expenditures, often as gifts at Christmas and birthdays. Fancy boxes for children's candy added another layer to the commodity form and increased its price. While poorer children ate their penny toys out of waxed paper bags and then sought out another cent, the rich ate their candy and still had an object left to play with or to admire. Whimsical papier-mâché boxes shaped like cars, circus animals, log cabins, shoes, and pieces of fruit—"strategic component[s] in the symbolic struggle for social distinction" among the elite[52]—themselves became toys or, more likely, display objects, once a child had eaten the candy. The product as a whole lasted much longer than the time it took to swallow a gumdrop or nibble a piece of licorice.

Often, these boxes mimicked adult luxuries. A description in the *Confectioners' and Bakers' Gazette* in 1905 of a candy box shaped like an automobile reads like a mail order catalogue entry and emphasizes accessories echoing the "toys" of the upper classes: "It has a fine plush seat and a polished brass horn, with imitation rubber bulb which every one who sees it is tempted to squeeze. It has a little electric lamp in front and its number in the rear. The three inch metal wheels are painted with imitation rubber tires." And so on. No mere box, the novelty container complete with working lights and imitation rubber tires was a miniature version of the most fashionable car around. A possession for the rich child, it imitated the luxurious toys of his parents and was designed by a "well-known toy manufacturer," whose "next specialty in this line will be a candy box phonograph."[53]

The verisimilitude of boxes such as cars and phonographs activated a world of play close to real life—a good life delivered via the technological ideal—for those who could afford it. As Susan Stewart has said, "once the toy becomes animated, it initiates another world, the world of the daydream."[54] Yet poorer children, by reason of their economic circumstances, lacked access to more realistic versions of the daydream; if they did not possess a penny, they were not even allowed into the store that offered the goods that initiated the daydream. And they certainly could not afford such luxuries as a fancy box. Even if these boxes were adult conceits used as gift containers to be admired rather than played with, they still only graced upper-class curio cabinets; but their presence in candy shops surely did not go unnoticed by children with their pennies. Children born into relatively affluent families who received elaborately packaged candies on special occasions would as adults buy fancy boxed French chocolates for their paramours.

A seemingly "innocent" place like the candy shop embodied the economic disparities that separated the classes and taught children the consequences of not having enough. A single penny could yield an array of things if channeled through the proper transaction. The possessions of the economically disenfranchised, though, lasted only momentarily—money spent on candy quickly vanished down one's throat. Like rent payments, these expenditures were not capital investments but bought only something fleeting and elusive. In contrast, richer kids, like their parents, owned concrete, enduring, tangible goods and actively engaged in the process of ownership through their candy box toys and prize premiums.

Inundated with such goods year after year, children gained a sense of aesthetic leaning and a material ethos through the nature of these objects. Most candy boxes came from Germany and France, because the United States lacked the artisanal structure to support the manufacture in quantity of such fine handmade goods. That the boxes were imported gave them immediate status as something exotic. That their shapes whimsically replicated familiar objects endowed them with yet another layer of meaning and value. In 1901, assigning import duties, even the Treasury Department determined that papier-mâché boxes transcended their functional purpose as containers—"filling things"—and "without doubt they are toys."[55]

Besides fancy candy boxes, manufacturers made glass containers that held candies. Shaped like boots, ships, elephants, and fruit, they accomplished what boxes did by providing a protective package that was a commodity in itself. Unlike the papier-mâché boxes, glass containers directly imitated a confectioner's display ethic by placing the treats within a transparent enclosure. They replicated in miniature the paradox of simultaneous accessibility and inaccessibility that characterized candies in a store, flamboyantly displayed in glass "exposition" jars that also rendered them untouchable.

The shapes of penny candies themselves reflected the increasing juvenile fixation on commodity forms. For example, George Mills's 1872 catalogue featured such candy shapes as allegorical figures, fountains, and people engaging in various activities. By the end of the century, the candy shapes became much more gender-differentiated and commodity-oriented, focusing on replicas of purchasable objects rather than on people's actions. The J. Frauenberger & Co. catalogue, which offered page after page of "Penny Cream" and "Penny Marshmallow" candy mold patterns, also demonstrates this trend. Its penny patterns included more masculine objects—guns, jackknives, wrenches, axes, policemen, soldiers, cigars, gin bottles—and fewer traditionally feminine objects—babies, purses, shoes, and cupids. That boys enjoyed more of a choice of boy-shaped candy toys presaged their teen spending habits. "Parents offered far less encouragement to girls to play at all, preferring them to spend their time in more useful activities such as sewing or minding the baby, or in quiet pastimes of embroidery or reading."[56] Public sites of immediate gratification—saloons, theaters, arcades—also remained male-dominated worlds. This may explain why girls, by early adolescence, purchased more durable goods, while boys still spent on immediate and ephemeral pleasures.[57]

At a very early age, girls took on household responsibilities that kept them tied to the domestic sphere.[58] The drop in their candy expenditures at adolescence did not mean a concurrent decrease in consumption, for teenage girls were often criticized because of how much candy they ate. Rather, girls went from buying candies as children to receiving gifts of candies as young women. In contrast, boys switched from one indulgence to another, becoming avid cigarette smokers as young adults. In fact, from the mid 1880s to World War I, reformers attacked cigarette smoking and blamed its origins on the candy habit. While 13-year-old girls bought "plant slips" with their money, boys of the same age spent their cash "half for tobacco and the rest for fun."[59]

⁖ *The Voices of Reform* ⁖

By the end of the century, candy-eating had become so universal among children that the social reformers of the day grew concerned. Physicians, domestic scientists, social workers, religious advocates, and other commentators all found something to criticize in candy, and they based many of their concerns on the fact that children could and did buy it without parental permission.[60] Changing over time, reform rhetoric expressed contemporary anxieties about freedom and indulgence. While in reality consumers by and large ignored reformers' admonitions (or at least the children did, and their parents lacked the ability to do anything about it), it is worth noting that many of these arguments expressed classist views that targeted the lifestyles of the working class and the poor. Elite reformers' pleas, although seemingly well-intentioned, attacked the social habits of those less fortunate and concentrated on issues relating to pleasure and autonomy and general social control.

Commentators frequently charged that penny candies were adulterated with harmful ingredients. Indeed, the local candy man did sometimes resort to using toxic dyes for coloring and adding other agents to extend his sugar supply. While there is no clear evidence that the composition of penny candies ever made children sick (although overeating certainly did), physicians and scientists alike were preoccupied with studying and publicizing the deleterious aspects of cheap candies. Although commentaries most often targeted penny candies, which represented the very bottom of the confectionery hierarchy, finer and imported confections were themselves often made with adulterated ingredients. As early as 1830, Dr. W. B. O'Shaughnessy reported in the *Lancet* that popular candies on the London market, many of which were exported to America, contained a wide array of adulterants, including red oxide of lead, chromate of lead, and red sulphuret of mercury. He also found that the inks used to print candy wrappers contained poisonous dyes, and that "*children invariably will suck or eat these papers,* from which it is evident the most fatal accidents may occur."[61] American confectioners faced the same problems years later, and a confectionery manual of 1866 warned: "Glazed paper, both white and colored, is often prepared with poisonous substances, and should never be used, as the candies frequently adhere to the paper, the particles of which may prove injurious." Children risked "sore mouths or inflamed gums by sucking or chewing these papers."[62] Even the most conscientious confectioners who prided themselves on selling goods with integrity could not guarantee that all their goods were pure. They especially ran a risk when relying on imported goods; a trade journal of 1901, for example, reported instances of powdered glass used in "high-class French" confections, to "give a glittering" to them.[63] The rich, it seems, purchased only the illusion of purity by spending more on fancy confections.

Yet the most strident critiques continued to focus on the lowly penny candy, the cheapest and most ubiquitous form of confectionery, and the one that most often reached the mouths of "innocents," because "children and young persons of both sexes, and especially girls, eat confectionery so largely."[64] Frank De Puy warned mothers, in his *New Century Home Book* of 1902, to "keep a close watch upon the 'penny candy' so alluring to most children. . . . Plaster of Paris and terra alba are often found in candy, and glue is used instead of gum arabic in some of the common gumdrops."[65] Adulterated candies endangered the very fiber of the domestic sphere, and they were even more insidious because their relative quality could not be judged by appearance alone. Christine Frederick, a noted domestic scientist, published an alarming account of twelve different local confectioners in a western town selling a "beautiful (!) assortment" of "penny line candies," all adulterated—"varnished, brightly colored with dyes, flavored with artificial products." People, she noted, "would not believe that such candies were purchased only two blocks from *their* homes!—or that such vile candies *were made!*"[66] Frenzied accounts such as this appeared with regularity in nineteenth-century popular literature and enumerated the many dangers, some literally around the corner, that threatened the well-being of increasingly independent children.

Focusing on issues of inner character, the reformer Bronson Alcott stridently warned of the dangers of confectionery, which endangered the spirit even more than the body. He believed in a direct connection between the "head and the genitals" and saw the intake of "fancier foods" as an act threatening one's moral fiber. Like sex, candy was a tempting indulgence, and it was thought to quicken the blood and incite impure thoughts. Alcott promoted fresh fruits and vegetables as safe alternatives to what he called the "demoralizing tendency of confectionary."[67] Reformers continued to voice their concerns about sugar consumption throughout the rest of the century, presenting a counterpoint to the advertising rhetoric aimed at children.

Because children could not be trusted to do what was good for them, parents and social advisors alike worried about their consumption of candies, and not just because the sweets might be adulterated. Some people believed that the consumption of even "pure" candies caused harm to one's physical and moral fiber and led to more deleterious vices. Adults also worried because candies motivated children to act like grown-ups. By purchasing their own goods, and using them to imitate adults, children blurred the lines between adult and youngster, between autonomy and dependence: "Children with their own money to spend could do as they pleased away from home,

FIGURE 11. Like father, like son. Boys in particular enjoyed candy goods that allowed them to act like grown-ups. Chocolate Cigarettes. Hawley and Hoops, New York. Trade card, ca. 1890. Warshaw Collection of Business Americana. Archives Center, National Museum of American History, Smithsonian Institution.

could consume or hide the evidence, and no one, least of all their parents, would be the wiser."[68]

Reformers and parents blamed early candy consumption for, among other things, leading to smoking. They had every reason to believe this, especially since confectioners did little to dispel these notions. Two popular forms of penny candies and penny chocolates mimicked cigars and cigarettes (fig. 11). Boys more than girls liked these candies, because they helped them imitate their fathers. "Boys had fun buying licorice, because they could emulate their elders by pretending they were chewing tobacco," one man recalled.[69] "Boys chew many such things as coffee and licorice to make their spittle look like tobacco in imitation of men who chew tobacco," Sanford Bell observed in a 1904 study of the "Psychology of Foods."[70] Bell saw this as a healthy impulse, because it helped "give additional experience to the sense of taste" (whatever that meant) and counted as a crucial part of the developmental process. But the candy cigar and candy cigarette remained a vexing combination for reformers, who did not like the idea that the candies poor and working-class children ate might inculcate more harmful vices.

Those who ran combination tobacco-candy stores peddled vices to all ages and watched their male customers grow from boys sucking on candy cigars to young men smoking real cigarettes. Candy cigars, one of the most common forms of penny candy, provided an easy and less harmful way a boy could imitate a man. It must have been an easy switch from candy to tobacco, especially when the second was offered within such close proximity to the first. In fact, so many underage children bought tobacco products that city governments slapped merchants with fines, against the objections of reputable cigar shops. After New York prohibited the sale of tobacco to children under sixteen in 1889, one store owner complained, "There are numbers of small dealers, candy stores, and news stores who have sold cigarettes and tobaccos to boys of any age so long as they brought the necessary cash. . . . The small boy will undoubtedly manage to procure his smoking material by some means or another, and with this the trade cannot cope."[71] Another man, fined for selling tobacco to a boy he knew was buying cigarettes for his father, also defended his position as a reputable dealer and suggested the people seeking the real purveyors of illegal substances look elsewhere: "I would like to see the candy stores touched up in this racket by our zealous police force."[72] The merchandising techniques used for selling cheap candies to children also promoted other "penny" goods like tobacco products. As many as ten or twenty cigarettes could be had for a nickel.[73] In both Britain and America, they were sold in "penny packets" and contained "tabs," picture card premiums especially alluring to boys. Almost identical to those found in candy packages, these serialized picture cards—of sports figures, movie stars, international flags, songbirds, and so on—prompted children to make future purchases so that they could "collect them all."[74]

That people noticed boys smoking tobacco in public made them feel even more adult. As early as 1867, two British travelers remarked on the precociousness of American children, who seemed like miniature adults: "It is not uncommon to see children of ten calling for liquor at the bar, or puffing a cigar in the streets."[75]

Stealing and gambling joined smoking as vices that reformers believed stemmed from an early candy habit. The American Sunday School Union used candy to relay a moral message about stealing in a lithograph from the 1860s entitled "The Young Thieves." The illustration shows children around a candy stand, considering their choices and purchasing their treats. In the meantime, two boys slink off to the side, looking furtively back toward the main scene. The accompanying text explains the scam, in which the two "rogues" sneak off with someone else's purchases. Relatively speaking, it is a small offense for small people, but is a theft nonetheless: "the evil which they have done to themselves is as great as if they had taken a pocket-book or a watch." "Having sinned, they yielded to the temptation of the devil. They have followed the evil devices and desires of their wicked hearts; and they are nothing more nor less than young thieves."[76] A convenient metaphorical trope, candy-stealing succinctly conveyed reformers' messages. Children related to the context because the candy man was a familiar sight and because his candy was tempting and easy to steal. Much later, in a 1905 study of boys' gangs, J. Adams Puffer found that a popular "Predatory Activity" was stealing, and that thirty incidents out of forty-nine reported robberies involved "fruit, cakes, candy, beer, and milk," to eat and drink, while another fourteen of the forty-nine involved stealing things to be resold, which also included fruits and candy.[77]

Children also resorted to gambling in order to increase their material possessions; this vice, too, had connections to candy. Richer children participated in sanctioned games of chance, like the "contests" and the mystery prize packages that confectioners offered. But turn-of-the-century candy stores also used overt forms of gaming to entice lower-class children. There were "kiddie slot machines" whose jackpots paid out in tiny pieces of gum. There were weekly lotteries awarding boxes of candy. "In Chicago, the Juvenile Protective Agency claimed that the gambling games in the candy shops had become so popular that children in one school 'were pawning their school books in order to get money with which to play,'" David Nasaw notes.[78]

Perhaps inspired by what they saw in the candy shops, the children who found their entertainment in the streets initiated their own games of chance involving penny candies and rigged roulette wheels. In some cases they used candy as the game and in others as the reward. Candy remained the constant element—a familiar and understandable form of juvenile cultural currency. The *Soda Fountain* published an article in 1906 entitled "Candy Roulette Wheels Tempted Children," which encapsulated reformers' concerns to date. A 24-year-old and a 26-year-old, charged with "running gambling devices to catch the trade of school children," combined the allure of candy (gluttony) with the allure of money (greed). Each man "had a rude roulette wheel made out of the top of a barrel, with compartments in which candy was placed. An arrow of wood was spun on a nail in the center. . . . patrons invariably lost, and received a very small amount of candy to pacify them when necessary."[79] The accused paid out just enough candy to keep children happy and to ensure their continued patronage.

What may have been construed as child's play followed some youths into their teen years. The *Confectioner and Baker* reported that one high school put up a notice warning: "Any young women found guilty of shaking dice or otherwise engaging in games of chance for the purpose of seeing who will pay for the ice cream soda or marshmallows in the stores surrounding the high school grounds will be immediately and summarily dismissed." While the girls saw the practice as being "perfectly harmless," for they played for only the price of a soda or candy, the "authorities" had "determined to stop the practice for fear that it may develop a love of gambling."[80]

Yet neither stealing nor gambling was the worst vice identified by reformers as an after-effect of youthful candy consumption. Temperance advocates, believing that eating candy led to alcoholism, had decried the practice since the 1840s. Popular commentators noted that sugar could be just as addictive as alcohol, and that children certainly could not control a desire for either thing. "The use of sugar is the stepping-stone to intemperance," James Redfield declared in 1866. Like the drunkard, he elaborated, the candy eater cannot resist temptation, and "requires larger and still larger potations of sugar to satisfy him," and "almost rushes in search of it, and goes from one candy-shop to another as the toper goes from one coffeehouse to another to satisfy himself with drams. . . . The appetite for the result of the saccharine fermentation is like that for the result of the vinous."[81] Reformers saw a logical progression of vice and postulated that the early taste for sugar led directly and naturally to a taste for liquor. To want these things came from the same impulse; the object simply changed with age from candy to the cocktail.

Candy manufacturers—who also continued to produce candy cigars and cigarettes—did little to allay temperance workers' fears on this account either. In fact, some of their advertisements even played up the potentially physiological addiction to sugar. For example, a trade card for Kandi Kubes, a popular turn-of-the-century confection, shows a boy eating a Kandi Kube with one hand while clutching a full bag of them in his other hand. The accompanying ditty reads:

> Handy-Spandy, Jack-A-Dandy,
> Wants a piece of PEANUT CANDY.
> He wants it now,
> He wants it bad.
> And KANDI KUBES just make him glad.

That he wants the candy "now" and "wants it bad," suggests that the immediacy of his urges can be satiated only by eating Kandi Kubes, which slake his thirst for candy and "make him glad." This refrain recalled the concerns voiced in *The Colored American* of 1837, which blamed candy for "creat[ing] appetites and tastes which cry give! give! give! and yet are never satisfied."[82] Similarly, Ned Higgins, Hawthorne's avid consumer of sweets, is so driven to obtain his ginger cakes when the shop is closed that, by the end of the story, he attempts to break a window: "There being no answer to several repetitions of the summons, Ned began to grow impatient; and his little pot of passion quickly boiling over, he picked up a stone, with a naughty purpose to fling it through the window; at the same time blubbering and sputtering with wrath." Hawthorne describes him as "stamping, shaking his head violently, making deprecatory gestures with both hands, and shouting . . . at mouth-wide screech."[83]

More literally, some candies even came in the shapes of liquor bottles, labeled "gin," and others concealed liqueur centers. A temperance crusader in the 1880s assailed "brandy-drops and rummy hearted caramels, declaring . . . that they would implant in the juvenile consumers of the syrupy *bon-bons* a taste for ardent spirits."[84] Others agreed that the use of liqueurs in certain candies constituted a terrible practice: "This must be regarded as an adulteration of a very reprehensible character, since these products are eaten so much by children and the danger of injury from the alcohol and the danger of forming a habit from eating it in this way is extremely great."[85]

Still others told stories of liquor being sold not in candies but alongside them. Chicago confectioners were said to sell cocktails as well. The Reverend R. A. White "told of seeing young girls sitting at tables in the place in question with gin fizzes and cocktails in front of them and with their minds in a befuddled state produced by drinking." White worried that these "young girls were being led from paths of decency and virtue" by alcohol's close proximity to the candy to which they had initially been attracted. *Confectioner and Baker,* the trade journal reporting this, countered the claims, calling White's tactics "sensational," and evidence of "yellow ministerial perversity." He might have found one example of such practice, but that in no way implied that all of Chicago's candy stores sold cocktails.[86]

Temperance workers kept stressing a connection between liquor and candy. In 1906, a writer in the *Ladies' Home Journal* invoked scientific and moral theories that warned women against letting their children eat candy, saying, "The first craving from ill feeding calls for sugar; later for salt; then tea, coffee; then tobacco; then such fermented beverages as wine and beer; and lastly alcohol."[87] A Dr. A. C. Abbott claimed in 1907 that "the appetite for alcohol and the appetite for candy are fundamentally the same, the choice of the one or the other indulgence being determined by the temperamental qualities of the individual."[88] Reformers could do little, however, to stop the production and consumption of confectionery, which continued unabated.

Without long-term studies of the effects of candy in the marketplace, however, people really had no idea what future physiological or mental consequences might result from childhood consumption, so they predicted the worst based on experience with other pleasurable consumables. Moreover, as David Nasaw points out, the arguments surrounding candy consumption were in many ways really debates about the larger place of children in society: "By isolating [children] from the environments that fostered precocity, the reformers would succeed in returning them to the status of full-time children, protected morning, noon, and evening from the temptations, the excitement, the sounds and sights of urban life."[89]

Even though writers of popular literature railed against the wholesale use of sugar, they covertly sanctioned middle- and upper-class consumption, indicative of the entire reform movement's selective permissiveness. Candy-eating by poor, purportedly innocent children could lead to all kinds of future sins and personality defects. However, the privileged accessed sugar in a way that absolved them of guilt and was ignored by reformers, because it was better disguised. In fact, the same forums that warned of sugar's deleterious effects—including women's magazines, home entertainment manuals, and cookbooks—also encouraged its consumption, the only difference being the form the sugar took and who consumed it. Candy *qua* candy—unwrapped, cheap, eaten in public—represented the uncontrollable nature of city life and a kind of contamination reformers targeted. In contrast, sugar that was "civilized"—eaten in the home and commodified through integration in elaborate menus or festive parties—represented one of life's pleasant, innocent luxuries.

For example, in 1873, *The New Cyclopedia of Domestic Economy* listed "species of food" not recommended for children, including pastry and confectionery, but on the very next page recommended that "To Prepare Fruit for Children," fresh fruit should be sprinkled liberally with sugar and then the whole mixture cooked to a syrup.[90] Concentrating it into an identifiable material form such as candy made sugar an isolated commodity that people easily linked to other indulgences and vices. Indeed, by shaping candies like gin bottles and cigarettes, candy makers took advantage of children's attraction to grown-up vices. But sugar itself trickled into upper-class diets in greater and greater amounts as an ingredient of cakes, sauces, syrups, preserves, and myriad packaged foods.

The economically able incorporated refined sugar into children's diets as a form of novel entertainment. *The Boston Cooking School Magazine,* a highly respected domestic economy periodical published at the turn of the century and aimed at the women of prosperous households, occasionally featured recipes for children to make. A typical example, "Recipes for Boys and Girls," appeared in the 1902–3 volume. Of eleven recipes, seven contained some form of sweetener (five called for regular sugar, one for maple sugar, and one for molasses), and six of the eleven were sweets: currant jelly; Swiss chocolate bread; chocolate and walnut canapés; maple, nut, and chocolate fudge; grape ice cream; and soft ginger cookies.[91]

Many children's cookbooks themselves emphasized sugar. *When Mother Lets Us Cook,* published in 1909, featured fifty-four recipes in all, of which thirty-two called for sugar. Twenty of the fifty-four recipes, or 37 percent, made specifically sweet treats. Many other children's cookery books followed the same pattern, with sweet outnumbering savory dishes. "Candies for Boys and Girls to Make" in the 1903–4 volume of *The Boston Cooking School Magazine* presented candy-making as an accessible children's pastime. The recipes emphasized "easy" candies to make—the same ones featured in domestic candy-making manuals—including three different kinds of fudge, penuchie (a maplelike fudge), creams, and simple bonbons.[92] Sugar had finally become such a common staple that children could literally play around with it at home, which was not only permissible, but encouraged.

The glut of turn-of-the-century instructional material included information on children's parties, which were becoming all the rage. Only those whose parents could afford them ever had such parties. Predictably, they mimicked the customs of the parents themselves, and in some instances, they provided living tableaus for the enactment of mother's fantasies, becoming animate versions of the doll house. The parties, highly thematic and scripted, often involved making candy, eating candy, or both. A Washington's Birthday Party, for example, featured "Individual George and Martha Washington ice-cream forms, cake with 'Mount Vernon' in frosting" which "may be procured from the best caterers." Red, white, and blue "George Washington hats filled with bonbons, cupcakes with white fondant-icing" added to the themed decor. After the meal, "The children, having partaken bountifully of the goodies prepared for them," then went to the parlor, where each received a patriotic-themed bag "filled with candy and popcorn."[93]

This profusion of sweets in myriad forms and in a succession of "courses," commonly prescribed in genteel women's magazines and domestic manuals, counted as popular entertainment and not as the first step down the road of irredeemable vice. Other "novel fads" for children's parties, equally championed by domestic scientists, included constructing maple-sugar log houses and eating ice cream in various forms. Parties centered around candy-making were also common. Fudge-making and taffy-pulling events occurred quite regularly among the elite. "'Candy pulls' are among the most enjoyable events of children's lives," proclaimed the author of one entertainment manual.[94] Another kind of party, reminiscent of the commerce surrounding penny candies, enacted the merchant-customer relationship. Each child received pieces of "tin money" on his arrival to use for purchases. "An older boy acted as merchant, and a variety of inexpensive toys were found on the counter for sale. There were whistles without number, perfume sachets, cheap jewelry, balls, and a variety of other toys, besides candy."[95] Candies, toys, treats, and games all merged in party form, where children pretended to be buyers and sellers, practicing play transactions in make-believe that mirrored the world just outside their rarefied domestic realm.

These parties created the guilt-free, transgression-free alibis that allowed upper- and middle-class children to eat sugar. In hard candy form, sugar could endanger one's health and threaten the purity of one's soul by offering great temptations. But dressed up in party form, sugar numbered among the many innocent pleasures of youth, which were encouraged by parents. Parents of privilege literally allowed their children to have their cake and eat it too, to enjoy the frivolities of theme parties and fancy candy boxes, which constituted suitable "packaging" for sugar. Poorer children, however, satisfied themselves with sugar in its more obvious—vulgar—forms, like penny candies, which came wrapped in waxed paper or not packaged at all, except in one's sweaty hand or linty pocket.

While the sanitized version of sugar consumption met with the approval of parents and society, the unsanitized versions, as reformers always pointed out, could lead to illness, or more gravely, to a future of perpetual transgressions ended only by a final fall into the abyss of evil. The rules about sugar consumption and children, while aimed mainly at the children themselves, were also directed at lower-class mothers. "The mother who keeps her bantling 'good,' while she talks or works, by relays of candy, more surely creates a craving which can bring no benefit and may work infinite evil," one writer admonished.[96] In other words, the poorer woman who could afford neither domestic help nor child care should not resort to pacifying her baby with candy, because she surely would be leading her child on the road to sinfulness.

It makes sense that the reformers' reactions against penny candies were particularly strident. These were the treats selected by children, and children put desires before safety concerns. Also, penny candies' ubiquity made them easy scapegoats for reformers searching to explain society's sinful ways. More to the point, penny candies brought simple and affordable pleasure, and their consumption evaded rules—even those imposed upon children by their parents. For that reason alone they became the targets of people bent on social control. The threat that sugar posed because of its pleasurable aspects was not confined to cheap candy but pervaded all forms of confectionery.

CHAPTER THREE

Cold Comforts

∴ *Ice Cream* ∴

In 1744, fourteen years after Nicholas Bayard's advertisement for sugar candy appeared in New York, William Black made a record in his diary of another confection, which he had encountered at a dinner given by Governor Bladen of Maryland in Annapolis. He described the meal as "plain proof of the Great plenty of the country, a Table in the most Splendent manner set out with Great variety of Dishes, all Serv'd up in the most Elegant way, after which came a Desert [*sic*] no less Curious, among the Rarities of which it was Compos'd, was some fine Ice Cream, which with the Straw-berries and Milk, eat most Deliciously."[1] Although this is the earliest known written account of ice cream embellishing an American dinner table, ice cream making must have been familiar enough for Governor Bladen to confidently serve it at his table to his guests—not the stuff of, nor the situation for, new experimentation. By the 1740s, many London confectioners were already serving ice cream, and it is likely that one of their personal recipes or a published one traveled to America and found its way to Governor Bladen's servants and eventually to his dinner table. Written accounts of ice cream in America do not appear again until the late eighteenth century, when even sweetmeats and comfits were still hard to come by. Freezing a liquid mixture laden with sugar, a luxury, by using ice, an ephemeral luxury, was something that elite Americans began enjoying only late in the eighteenth century and primarily through the institutional trappings of the urban pleasure garden, which established the way Americans came to perceive ice cream and the contexts for its consumption throughout the nineteenth century.

Ice History: The Cold, Hard Facts

Like those of sugar, the origins of ice as a commodity are impossible to pinpoint with certainty. Elizabeth David writes that as long ago as 1100 B.C., a Chinese poem referred to ice harvesting and storing. By A.D. 400, if not earlier, the domestic use of ice was common practice in Persia. And in T'ang dynasty China (A.D. 618 to 907), rulers used it in sophisticated ways—as summer air conditioning and for food preservation.[2]

In more recent history, the Florentine gentry of the late sixteenth century used their many icehouses not for food preservation but rather to store ice and snow that chilled beverages in the hot summer months. The *Libro de la nieve,* published in 1571 in Seville, instructed the elite on how to cool wine and other beverages—by nestling glass decanters in mounds of the stuff and spinning until the beverages they contained turned cold. By the 1660s, Italians enjoyed "sorbets," a kind of icy-slushy beverage, and Italian confectioners were making, along with grand pyramids of ice embedded with fruit for lavish celebrations, something called *candiero,* a semi-frozen mixture of eggs, cream, and sugar. Concurrently, Europeans experimented with using saltpeter as a refrigerant in order to make man-made ice. People treasured ice as they did sugar; a desirable commodity, it inspired experimentation and grand products.[3] Ice also, like sugar, signified power, wealth, and a command over the resources of both man and nature.

Ice did not remain a mere coolant for beverages but possessed spectacular culinary possibilities. Many French and Italian confectioners and *liqueuristes* (beverage men) published books in the late seventeenth century giving directions for making sculptural ice pyramids, water ices, and *cresme glacée,* perhaps the first proto–ice cream recipe to appear in print. Many confectioners plagiarized these recipes, and the English translation of one, published in 1702, inspired a wave of "iced creame" making and eating that remained relatively unchanged until Hannah Glasse's instructions for raspberry cream ice appeared in 1751.[4] The written documentation of recipes and their transmission through publication, the international couplings of royal families (who brought along personal status symbols to their new places of sovereignty), and of course the simultaneous spread of sugar cultivation and refining all coincided, spreading the fashion of ices and ice creams to many European countries at roughly the same time, fashions refined Americans happily imitated in the eighteenth and nineteenth centuries.

Originally, Europeans consumed ices and iced creams during private ceremonial affairs (if one can consider courtly banquets for hundreds of people "private"). French and Italian royalty, employing their own confectioners, served these ices on their own premises exclusively. Not until late seventeenth-century Paris did eating ices and iced creams become fashionable among anyone other than royalty, when people began consuming them in public venues. The French and Italians constituted the vanguard when it came to the imaginative use of ice. Francesco Procopio, a famous confectioner and *limonadier* working in Paris in the 1660s, came up with the idea "to decorate the walls of his renovated establishment with fine large mirrors, to hang crystal chandeliers from the ceiling, to provide marble tables at which customers could sit in relative comfort to enjoy their coffee and chocolate, their sweetmeats, their sticky, scented fruit syrups, and in the summer, those famous frozen sherbets and *neiges.*"[5] The original ultra-fancy contexts of consumption for ices and ice creams, both public and private, would persist. The kinds of material trappings detailed above uncannily resembled those found in exclusive American ice cream saloons of the early and mid nineteenth century, which very likely borrowed from the French aesthetic.

By the end of the eighteenth century, ices and iced creams "had become quite a craze in fashionable London." (That Governor Bladen served ice cream at his table in the 1740s certainly made him one of America's culinary avant-garde.) Domenico Negri, a transplanted Italian confectioner who became the "king-pin of London confectioners," offered syrups, candied fruits, biscuits, and other treats in addition to the cream ices for which he was renowned.[6] Frederic Nutt, an apprentice to Negri, first published his *Complete Confectioner,* a work that enjoyed popularity both in Britain and America among confectioners, private cooks, and bibliophiles alike, in 1789. First published in America in 1807, Nutt's book directly competed with the multiple British and American editions of Hannah Glasse's *Compleat Confectioner,* first published around 1760, which appeared in at least seven different editions in Britain alone. Both works included many recipes for ices and ice creams. Glasse's 1758 raspberry ice cream differed from the one for apricot ice cream she published in 1796, which appeared virtually unchanged in Elizabeth Raffald's 1801 Philadelphia edition of her *The Experienced English Housekeeper.* All required the same basic processes, and Glasse's recipe, from 1758, is typical:

> Take two Pewter Basons, one larger than the other; the inward one must have a close Cover, into which you are to put your Cream, and mix it with Raspberries, or whatever you like best, to give it a Flavour and a Colour. Sweeten it to your Palate; then cover it close, and set it into the larger Bason. Fill it with Ice, and a Handful of Salt; let it stand in this Ice three Quarters of an Hour, then uncover it, and stir the Cream well together; cover it close again, and let it stand half an Hour longer, after that turn it into a Plate. These Things are made by Pewterers.[7]

Printed information circulating in America via imported and domestically printed recipe books popularized novelties like ice cream not only by providing instructions for confectioners and private families but also by introducing the frozen treats into popular consciousness.

Fooling Nature: Ice Harvesting

People enjoyed eating ice cream because it tasted good, because it defied nature, and because it signified human ingenuity and power—the result of hard labor and a clever alchemy that fused the seasons with winter ice and summer fruit. In the era before mechanical refrigeration, people who came into contact with something frozen when the temperature outside sweltered enjoyed a novel and physically satisfying experience. Americans cut their ice from rural ponds and lakes rather than from nearby river accretions, which were often polluted with the waste products of processors operating on the banks, such as tanneries and abattoirs. Rural businessmen with the power not only to harness nature but also to control local supplies, ice dealers filled icehouses of those living in the hinterlands first, if they did not possess their own icehouses, and brought what remained to the city, where people paid considerably for this precious commodity. One of the few extant eighteenth-century ice advertisements, which appeared in the *Pennsylvania Gazette* on

13 July 1785, informed its readers that there was "ICE TO BE SOLD, At the Corner of Arch street and Sixth street, by the BUSHEL or HALF BUSHEL," a supply procured from the depths of an outlying icehouse, no doubt.

Effective icehouses of the eighteenth and nineteenth centuries stored ice by the ton (some were claimed to hold a hundred tons)[8] in a compact form and preferably underground, where they kept the mass frozen, dry, well-ventilated, and protected from heat sources. The prototypical icehouse consisted of a tapered pit dug twelve to sixteen feet into the ground, shored up and floored with planks of timber (which also allowed run-off drainage) and lined with straw to keep the interior dry. A small cylindrical building with a conical thatched roof to divert falling rainwater stood over the pit to protect the contents and allow for access, unnecessary if the storage chamber was dug into the side of a hill. By 1800, American icehouses were common and effective enough for their contents to be used not only to preserve fresh meat but also to cool drinks and make fancy confections.

J. B. Bordley, who wrote about Chesapeake icehouses in his *Essays and Notes on Husbandry and Rural Affairs* of 1799, told of a tavern icehouse he saw in 1797, "on Gloster [*sic*] point near Philadelphia. . . . It was then full of ice, in pieces the size of apples." Bordley returned a year later to again inspect this icehouse and "was assured that the 61 loads kept through the summer, and that 'some loads of ice were in it when ice came again.'"[9] Indicating newly popular uses for ice, and suggesting a reason to build one's own icehouse, Bordley's treatise included a copy of Glasse's ice cream recipe of 1758 virtually unchanged, substituting strawberries for her raspberries. Other icehouse engineers included men like Thomas Moore, who published a treatise on refrigeration in 1803 explaining how to build a structure that would hold about ten tons of ice for less than $20, "if we calculate on a great part of the business being done by the family; which in the country in general, it very well may." Moore also experimented with a portable refrigerator which kept perishable food for regular households, for people going to and coming back from markets transporting "poultry, veal, lamb, and all sorts of small marketing," and for businessmen like butchers and fishermen.[10]

Ice stored in icehouses and used in contraptions like Moore's refrigerators came from hoarded winter supplies, which men harvested like many other natural crops. Farmers, "who hacked and sawed chunks from near-by ponds," supplied ice to city ice depots. And hospitals, putting ice to medical uses, remained loyal ice customers and sometimes kept their own icehouses.[11] Ice used for both practical and trifling purposes made ice harvesting and dealing into a profitable pursuit.

Clearly, even the first step toward making ice cream, procuring the ice, constituted an arduous and time-consuming task in the eighteenth and nineteenth centuries. Actually using it to make ice cream involved the combination of many costly items beyond ice and sugar; it was a rare treat for the elite even some sixty years after William Black's pleasant experience with Governor Bladen. Ice cream required sugar, cream, eggs, flavoring, ice, salt, metal buckets, and copper molds. These things, plus the time and labor needed to churn the constituent "tea" while freezing to guarantee a smoothly textured result, designated ice cream making and eating only for the very rich. George Washington's French confectioner made it for the president, Dolley Madison served it at state dinners, and Thomas Jefferson included experiments with it in his personal recipe book.

Even after ice cream had become an established marketplace sight, democratized about the same time as penny candies, in the 1840s, it retained its associations with wealth and luxury. While the packaging of some later confections added to their value as consumer commodities with certain kinds of social cachet, ice cream defied physical packaging until the advent of cardboard cartons in the twentieth century. Instead, what counted as ice cream's "packaging" proved more elusive. In effect, the physical locations in which people ate ice cream constituted its packaging and created its commercial aura. Sites of consumption denoted the status of commodity and consumers alike.

Pleasure Gardens

Early American politicians reinforced ice cream's high regard by featuring it at lavish, important dinners; they quickly went from coveting it in private to celebrating it in public. Ice cream first appeared in public at pleasure gardens, where the frozen treat became a requisite part of the "entertainments" provided. Popular in major American cities from the 1780s to the 1830s, these urban idylls were eventually superseded by ice cream saloons and yet later by ice cream soda fountains.

Owners of American pleasure gardens modeled their enterprises on British versions like the Tivoli and Vauxhall. Ice cream proved such an integral element to these places that early ice dealers tried to ship their supplies even to pleasure gardens located in European colonies to satiate the fashion-conscious desires of the planters there. In 1806, Frederic Tudor, a Boston ice dealer, approached a pleasure garden proprietor on Martinique about buying his ice. "Determined to spare no pains to convince these people that they can not only have ice but all the luxuries arising as well here as elsewhere," Tudor went about using his ice to make ice cream. He convinced the skeptics by selling $300 worth of ice cream at one saloon on his very first night trying.[12] Tudor hoped that the established taste culture of the expatriates would inspire a market for such luxury goods on the islands. Rather than promoting ice for chilled beverages, the Boston entrepreneur appealed to their desire for ice cream with the knowledge of its necessary presence at the pleasure garden.

The American pleasure garden, like its British counterpart, remained an urban institution that provided relaxation and respite from the urban pace in the form of a luxuriant park. People flocked to pleasure gardens during the summer in search of escape and entertainment. Refreshments, including ice cream and ices, and "amusements"—anything from string bands to nighttime illuminations—were essential to the pleasure gardens' allure. The gardens created worlds within fantastic worlds available to those with disposable time and money. Like people in other eastern cities, Philadelphians supported many such gardens simultaneously (at one time over ten), including Tivoli Garden and Gray's Ferry and Gardens.

Vauxhall, named after the famous London garden, opened in 1814. It is no wonder that visitors called it a "little paradise." Typical accounts include glowing descriptions of the surroundings: "There was always good music, the turtle soup and the ice cream were of unexceptionable [unparalleled] quality, and part of the illumination consisted of vari-coloured lights hung among the boughs of the trees. . . . From time to time there were elaborate fireworks depicting, for instance, such subjects as Niagara Falls." Owners shaped their gardens to conform to the tastes of the people visiting—those belonging to the class that had the time and money to spend in and on such a wonderful escape. Pleasure garden proprietors felt it incumbent upon them to provide novel entertainments in line with the fantasy atmosphere they projected, and that would live up to their promise of being magical worlds away from people's real ones. One observer at Gray's Gardens in Philadelphia wrote, "I fancied myself on enchanted ground, and could hardly help looking out for flying dragons, magic castles, little Fairies, Knight-errants, distressed Ladies, and all the apparatus of eastern fable."[13] The physical space, which combined exotic plants, long promenades, shade trees, and fountains, helped create and sustain these magical worlds.

Ice cream's cool and enchanting flavor found a fitting place among the many pagodas, temples, exotic plants, bridges, and grottoes. A heavenly substance—cool, sweet, and light—ice cream belonged to and in the pleasure gardens as a form of human ambrosia. It was logical, then, that confectioners played a large role in the creation and operation of urban pleasure gardens. For example, the Philadelphia confectioner and distiller Laurence Astolfi opened his Columbian Garden in 1813 and after the first season operated it without theater performances, solely as a refreshment garden. Although Astolfi's newspaper ads during the summer of that year promised impressive fireworks displays, they were illustrated with a cut of ice cream in a tall parfait glass, representing the typical serving size of the time, which was about two tablespoons. Philadelphia's Vauxhall, opened in 1814 as a "fashionable center," likewise showcased the works of its principal owners, at first an Italian perfumer and hairdresser named John Scotti and later a confectioner named Joseph Letourno. An advertisement appearing in the June 11, 1819, issue of the *Aurora* for the garden reprised Astolfi's advertising strategy, showing a similar illustration of a glass of ice cream rather than any images of the garden itself (fig. 12).

Pleasure gardens, not places for the common man, admitted the genteel only. Some enforced their exclusivity by requiring admission fees outright, while others, located in remote areas, stood far away from the pedestrian traffic of the manual laborer. For example, John Scotti charged 50 cents for entrance into his Philadelphia Vauxhall in 1816 but promised that "25 cents will be returned in refreshments of the best quality," including, "All sorts of Ice Cream, Sorbet, Cold Sabovoni, Pastry and Sweetmeats, Wines and Liquors of the first quality." Vauxhall offered "all that can be desired for Breakfast, Dinner, and Colations Cold," and gave its leftovers to the poor.[14] Reformers later in the century applied ideals from the pleasure gardens to public parks, which they felt improved the well-being of urban workers, but patrons of the early pleasure gardens flocked to them because they excluded common riffraff.

Confectioner-owners displayed their wares on both large and small scales. Because inns and taverns lacked room for outdoor amusements other than games like ninepins, the proprietors of pleasure gardens in Philadelphia and

Camden, *June* 11*th*, 1819.

VAUXHALL GARDEN.

THE PROPRIETOR OF THIS GARDEN,

MR. JOSEPH LETOURNO,

HAS the honor to return his thanks for the many favors which he has received, from the Ladies and Gentlemen of the city of Philadelphia and Camden, and he hopes by strict attention, to merit the patronage of a generous public. He informs them that the Garden has been fitted up, this spring, in a very neat and handsome style.

Ladies and Gentlemen can be accommodated with the best COFFEE, CHOCOLATE, TEA, and all sorts of RELISHES, including OYSTERS, cooked to order; also, CAKES and ICE CREAMS of every kind — DINNERS for parties will be served up at any hour, at a short notice.

He also informs them that the Steam Boat *Manette*, has commenced running from Market street, Camden, to Hepperd's ferry, a few doors above Market street, Philadelphia, and will continue running every day until 10 o'clock in the evening.

Those who wish to avail themselves of the advantage of *quarterly tickets*, either single or for their families, will please to apply at the bar in the garden, or at Spark's, on the wharf, or the captain on board.

Single ferriage at 6¼ cents.

June 16 d10t

FIGURE 12. Ice cream, more than the landscape or clientele, drew people to the early pleasure gardens. "Vauxhall Garden," *Aurora,* 11 June 1819. Library Company of Philadelphia.

other cities found their own profitable niche. They attracted more business in the summer by providing alternatives to saloon fare, such as "ice cream and other cates." In addition, they expanded both their entertainments and locations to accommodate more people, and "the pastry-cook in time became the astute, experienced showman."[15] Much of a confectioner's work operated on an aesthetic level, making a good appearance of utmost importance. The gardens enabled artistically inclined entrepreneurs to be highly successful in their pursuits, because these places were, in essence, confections on a grand scale. Topiary and shady bowers emphasized play, amusement, freedom, escape, and a lavishness also found in popular confections such as ice cream and sugar sculptures.

People sought sociability and escape at pleasure gardens in their various manifestations. Not surprisingly, people "escaping" the urban environment actually desired places where they could simultaneously meet their genteel peers and avoid the chaos of city streets. To these ends, pleasure gardens imposed regulations. At Scotti's Vauxhall, for example, "Children will not be admitted but in company with their parents," and "Strict attention shall be paid to the observance of decency and good behavior."[16] In addition, different pleasure gardens offered different kinds of diversions and enticed different gatherings of people. For example, Baltimore's five pleasure gardens operating at the end of the eighteenth century each catered to a discrete audience. Some gardens were sports-oriented and allowed only men. Others encouraged the patronage of women and children by promising an environment of "politeness, delicacy, and uniform conviviality," and the absence of "unprincipled fellows," who might disrupt the garden's serenity. The more formal gardens observed strict etiquette: "Women came in full evening dress, and men walked bareheaded, with their hats under their arms."[17]

Pleasure gardens created social spaces where people went to meet others, to discuss business, and to debate politics. Most illustrations of eighteenth- and nineteenth-century pleasure gardens show crowds of people enjoying sporting activities, walking through stands of shady trees, or gathered at tables imbibing refreshments. Barbara Sarudy, writing chiefly about Baltimore's pleasure gardens, suggests that the pleasure gardens' lush and remote atmosphere with plenty of shade trees and light refreshments also encouraged romantic encounters between the sexes. A New York state tourist's guide of the 1870s retrospectively described one such pleasure garden, Elysian Fields, as "*once* the famous and romantic resort of lovers," and city mysteries often described pleasure gardens as places where prostitutes loitered. Owners maintained this sense of separateness by requiring an admission fee per visit, or by offering subscription tickets for entire seasons, so patrons could be assured of meeting "no ungenteel people."[18]

Along with oysters "cooked to order," cakes, coffee, chocolate (to drink), and sometimes alcohol, the successful pleasure garden proprietor offered ice cream. It was one of a select group of comestibles that represented conviviality and indicated a venue's relative degree of refinement. Pleasure gardens introduced the genteel to ice cream, and even though they admitted only the elite and comfortable, local newspaper accounts of and advertisements for these retreats made it clear to all who could read, or even see, that ice cream counted as a grand treat.

Pleasure gardens symbolized man's differentiation from and triumph over nature. Exotic plants, including orange trees and pineapples, and reproductions of Chinese temples communicated this ideal. Illuminations at night contradicted the movement of the sun. Sculpted foliage mocked the wayward growth of natural vegetation. Ice cream defied the cycle of the seasons by capturing the coolness of winter in the middle of the summer.

Urban pleasure gardens enjoyed their heyday from the 1780s to the late 1820s. What was meant as yet another spectacular entertainment at Philadelphia's Vauxhall in 1819 turned into a disaster when observers grew impatient awaiting a balloon ascension. Both people who had paid their dollar admission and those standing outside the gates charged the balloonist, "and an attack upon the Garden followed. The fence was torn down, the balloon ripped open, the stock of wires and liquor looted, and the theatre building set on fire. In fifteen minutes, beautiful Vauxhall was a ruin."[19] The remnants of Vauxhall stood ignored for over four years. It was revived only during the 1824 and 1825 summer seasons, after which it became an empty lot, upon which a mansion was later built. Other pleasure gardens had gone the same way by the early 1830s. Urban expansion encroached upon these once out-of-the-way venues, making them no longer the attractive refuges they had been. What pleasure gardens had offered in terms of escape and respite was eventually diluted into the forms of the public park and the amusement park. Pleasure gardens that lingered on in various incarnations for the next few decades included Niblo's, which became a New York City theater, and Elysian Fields, which became a working-class park in the 1860s and a baseball field in the 1870s.[20]

After pleasure gardens as such disappeared, the refined found similarly exclusive places in which to eat their ice creams and other confections. To see the forms these public, class-based recreations took during the rest of the nineteenth century, we need only follow the path of ice cream, the epitome of refinement and gentility, which retreated indoors to ice cream saloons before the masses ventured outdoors to the public parks.

Although pleasure gardens had largely disappeared by the 1830s, this did not mean that the confectioners who ran them went out of business. Rather, they improved their individual shops and continued to produce fancy treats for those who could afford them. Confectioners' sense of display, playfulness, and amusement did not wane; they merely channeled it into smaller forms of delight for an equally exclusive clientele, who partook of confections on the semi-public level of the domestic dinner party or banquet rather than in the context of the pleasure garden.

Ice Cream Saloons and Confectioners

During the first decade of the nineteenth century, enough people made and ate ice cream to support "ice cream houses" or "saloons" in addition to pleasure gardens. Yet the novel substance remained enough of a curiosity to provoke controversy in the popular press. In 1809, for example, a writer in *Poulson's Daily Advertiser* using the pseudonym "Fornax" (Latin for furnace) not only accused ice cream of damaging teeth by juxtaposing hot and cold but went on to charge: "The other evening a young lady, coming out of an Ice Cream House, fainted, after having eat [*sic*] two or three glasses of this cold pernicious composition."[21] "Amatrix Refrigerantiae" replied in a letter to the editor the next day:

> I agree with your friend "Fornax" that it is highly pernicious to eat *three or more* Ice creams in succession when the body is heated; to say the least of it, it is highly imprudent. But Mr. Fornax is of the opinion that the loss of our teeth, at an early age, may be attributed to our drinking "hot and cold *luxuries,*" the conclusion he draws, I presume, is that ice creams are, in part, the cause of this defect in our beauty. . . . I am sure that

> "Mr. *Fornax*" would be severely punished, at this hot season of the year, if he were not permitted to cool his *furnace* by a draught of cold water or punch. . . . I am, however, sensible to his good intentions, in warning us against the use of *ice creams;* but I think, with due deference to him, that we ought to enjoy with *moderation* the luxuries with which it has pleased Providence to bless us. A lady fainting after having eaten *three ice creams* is no reason why *one* would injure me.[22]

The exchange articulated the many issues swirling around luxury, consumption, and sugar at the time, which continued to vex Americans throughout the nineteenth century. That its positive and negative qualities could be debated in a public forum (partially tongue-in-cheek) meant that ice cream already possessed some cultural importance. The (presumably) male writer's identification of women in particular as vulnerable consumers, and the female writer's defense of her gender's right to consume ice cream, proved an enduring trope, which recurred with increasing regularity in the ensuing century—women were especially vulnerable to sweets, and sweets had the potential to incite a woman's appetite beyond her control.

The availability of ice cream hinged on having access to the ingredients. Philadelphia, long known as the "Ice Cream City," was particularly fortunate in this respect. The nearby Schuylkill and Delaware rivers grew healthy ice crops each winter, and the fertile outlying areas abounded with fresh fruit, eggs, and cream, which farmers regularly delivered to the city. In 1835, one Philadelphian remarked on ice cream's growing popularity, recalling that the city had "not even [had] a respectable oyster cellar so recently as 1815," and that "ice-creams, those essentials now-a-days, were an expensive and rare luxury." Now, however, the writer elaborated, "for twenty-five cents a week you may enjoy the great luxury of Schuylkill water cooled with Schuylkill ice; as for ice-creams, they are sold in the market and everywhere, for a penny a glass, rivaling in cheapness the Penny Magazine, and a good deal cooler to take—we *would cut* any magazine for an ice-cream at any time during the dog days."[23]

If ice cream did sell for a penny a glass in 1835, it was certainly an anomaly, since the frozen treat did not stabilize at that price until some fifty years later, when it could be manufactured with cheap and artificial ingredients. During the 1830s, ice cream was only beginning to reach the palates of the middle classes. By the late 1830s, a glass of good ice cream sold for 12½ cents, half as much as the same two tablespoons had twenty years earlier; but this price still put it beyond the reach of the common man. In 1850, *Godey's Magazine and Lady's Book* proclaimed that ice cream had, oxymoronically, become one of life's "necessary luxuries," as crucial to a fancy occasion as a roast to a dinner.[24] Ice cream remained a status symbol because many people still could not afford the daily deliveries of ice, the time, or the labor to make this ephemeral confection.

Its physical and sensual qualities added to ice cream's value and also helped construct its social identity. Being cool, light-colored, sweet, and smooth, it embodied heavenly purity. Trade cards, for example, portrayed ice cream freezers as inhabiting the same fantastical worlds as dainty winged putti (fig. 13), influenced perhaps by earlier confectioners' books like Emy's *L'art de bien faire les glaces* of 1768,[25] which showed similar winged figures carrying ice cream skyward. Ice cream, especially on a hot day, tasted and felt like pure ambrosia: its flavor lingered on the tongue longer than the substance lasted in one's mouth.

Confectioners emphasized the cultural associations of ice cream by appealing to the elite's aesthetic sensibilities within established ice cream eateries. Their sense of display and showmanship found outlets in fashionable hotels of the time, exclusive summer resorts located in far-off places (like Cape May, New Jersey, during cholera epidemics), and within their own shops. Ice cream saloons, like their pleasure garden precursors, catered to the comfort and enjoyment of the upper classes. "Treating" their customers right, confectioners kept their clientele exclusive and loyal. Early ice cream saloons, places of luxury and indulgence, represented the same values as the pleasure gardens, but reconfigured them as indoor retreats in closer proximity to urban consumers (fig. 14).

Parkinson's Refreshment Saloon and Illuminated Gardens, the most famous, faced tony Chestnut Street in Philadelphia. It was opened in the early 1830s by George Parkinson, a well-known Philadelphia confectioner, whose family also concocted many of the recipes in the widely circulated book *The Complete Confectioner.* In "the most delightful lounge of its class in the union," Parkinson and his son James served all kinds of confections and also offered meals "in the most recherché style." The Parkinsons' amusements included illuminations and musicians who would "enliven each evening with a splendid programme of Operatic and Popular Music."[26] A miniature pleasure garden

FIGURE 13. Ice cream's physical properties associated it with winged figures who could ascend into the heavens. American Machine Company, Philadelphia. Trade cards, ca. 1890. Historical Society of Pennsylvania.

FIGURE 14. Higher-class ice cream saloons offered pleasure garden–like features, even in urban settings. W. S. Hedenberg's Ice Cream Garden, Newark. Ca. 1855. Courtesy of Harold and Joyce Screen.

in the city (described as a "cool and refreshing resort in the hot summer evenings"), Parkinson's borrowed heavily from those formerly popular places of respite and entertainment. "The garden attached . . . is filled with rare exotics, noble trees, shrubbery and flowers. A beautiful and classic fountain forms the centre, from which a *jet d'eau* is seen sparkling in the sunlight."[27]

George Parkinson ran his Philadelphia confectionery until 1838, when his son became a partner. The business consisted of two separate buildings, one housing the "manufactory" and the other displaying, serving, and selling the confections. No mere retail outlet, however, this operation provided a lavish place of comfort with different rooms to serve discrete groups of customers, its decor reminiscent of Procopio's marble- and crystal-laden Paris establishment of the 1660s. The "Gentleman's Room" "makes people feel fancy almost . . . that they are gentlemen, for it is a most gentlemanly thing to feed at Mr. Parkinson's on strawberries, with ice cream *powered* over them."[28] More than the ice cream, the space itself—forty-five feet deep, with an inlaid marble floor and murals on the ceiling executed by a locally famous fresco painter—was the saloon's main attraction. The saloon cost the elder Parkinson $30,000 in the 1830s and sported marble-topped rosewood tables, hanging mirrors, velvet carpets, and chandeliers.[29] Parkinson even installed "sofas, some ten or a dozen," to be "*layed on* by dandies." "There," one description continued, "one may *stare* and be almost satisfied, without calling out for anything." A lengthy paean to Parkinson's business published in a local newspaper, in fact, mentioned ice cream but twice. People frequented Parkinson's because of its lavish appointments and the owners' determination to create a genteel and elegant atmosphere: "Such luxury. [S]uch a convenient mode of getting rid of spare cash, not intended for the printer, we never saw, *till we went up Mr. Parkinson's stairs!* . . . Mr. Parkinson has set out, and does set out, the nicest tables and has rooms in the best taste, and the best tasted things imaginable."[30] Fine confectioners like Parkinson saw the inextricable link between physiological and cultural taste. To emphasize the air of exclusivity, Parkinson provided separate milieus for his customers, offering in addition to a room for both ladies and gentlemen, the exclusive "Gentleman's Room."

Other entrepreneurs, too, offered comfort and gentility to their patrons and aspired to Parkinson's standards of fashion. Like pleasure gardens, ice cream saloons were places for the refined to socialize, conduct business, or just relax. Ice cream, in fact, became a sort of euphemism suggesting the larger menu of "dainty" offerings. The presence of "fine ices" and "fancy creams" meant the presence of fine and fancy company as well. For example, Isaac Rodgers, another Philadelphia confectioner operating around the same time, charged 62½ cents per quart of ice cream in 1845, still a princely sum.[31] This kind of environment—a context public enough to encourage social intercourse yet private enough to maintain exclusivity—justified the high price of ice cream there and made it so popular among the comfortable classes.

Parkinson's Confectionery served as a prototype for ice cream saloons, which gained popularity in the 1840s and became culinary mainstays by the 1850s. Other saloons, usually smaller enterprises, were run by individuals who, like Parkinson, made their own confections on site but served them in more humble quarters. One man remarked in 1847 that "now it seems in some parts of the town as if every other house had been converted into a 'Creamery.'"[32] Fiercely competing with each other, these "creameries," or saloons, appropriated the vocabulary of Parkinson and his ilk, claiming gentility and exclusivity in order to gain customers even if they numbered among the seedier storefront operations of the day. Some even touted temperance principles by offering alternative libations to the ones found on tap in the beer saloons. Yet they reiterated the familiar promises of comfort, retreat, and special services.

By the 1850s, however, ice cream had become so common and affordable that the widely disparate contexts of consumption paralleled and highlighted economic divisions among customers. In his 1850 book *New York in Slices,* George Foster described some of these class-related distinctions. Ice cream saloons in Brooklyn and the Bowery signaled broader cultural disparities, but Foster also suggested that the trappings of the Broadway saloon had their own superficiality, being "generally fitted up in a style of exaggerated finery, which has a grand effect from the street, but is a little too glaring and crushing when you are within." Foster noted that the special effects of the Broadway saloons had even more impact at night, "when the gaudy curtains, silver paper, and gilded mirrors are highly illuminated by gaslight," and "the scene rises almost to splendor itself," or a version people accepted. Foster continued, "In the sultry summer evenings, every one of these fashionable Ice-cream Saloons is crowded with throngs of well-dressed men and women, belonging for the most part to the great middle classes; while the establishments in the Bowery are crammed to the very threshold with the b'hoys and their buxom and rosy sweethearts." Following its historical trajectory, the consumption space for ice cream remained a highly classed and gendered one. Foster noted these divisions, saying, "in day-time . . . the ice-cream saloons are mostly patronized by ladies, who stop on their way from Stewart's to regale themselves with a dish of fragrant souchon and a sandwich, or perhaps the more exquisite but dangerous luxury . . . of a sherry cobbler."[33]

Critics and commentators continued to remark on the feminine makeup of confectionery's key consuming group, which led to increasingly concrete stereotypes regarding the nature of women and the goods themselves. The Philadelphia *Sunday Dispatch* of September 2, 1855, for example, contained four different advertisements for ice cream rooms, ice cream saloons, and ice cream gardens, including one for the Ladies' Restaurant and Ice Cream Rooms, which boasted "the choicest delicacies of the season" and "Ice Creams, Jellies, Water Ices, &c., which for purity and richness of flavor, cannot be surpassed." Other high society ice cream parlors included Mr. Gohl's Confectionery and Ice Cream Saloon, and that of T. Slatter, Confectioner, who "solicits the patronage of the Ladies and Gentlemen of Philadelphia." Slatter, evoking the pleasure garden motif once again, promised "Ice Cream Gardens in the most modern style, with fountains throwing sparkling jets of water, and he assures the public that his Establishment will be kept in the most select manner."

Because they sold treats that some people still believed were potentially physically dangerous and morally deleterious, ice cream saloon owners repeatedly tried to defend their businesses as alternative retreats to the dark, smoky taverns selling alcohol. Sensational accounts appearing in popular literature linked ice cream with alcohol and paralleled reform movements against penny candies. In *Hot Corn,* published in 1854, Solon Robinson described the danger of such alluring places that sold alcoholic libations alongside the frozen ones. Of Taylor's ice cream saloon, he asked, "Was ever eating and drinking temptation more gorgeously fitted up? How the gilt and carving, and elaborate skill of the painter's art glitter in the more than sun-light splendor of a hundred sparkling gas-burners. . . . 'Tis the palace of luxury—'tis but a step from the palace of the tombs—'tis but a step beyond to the home of 'the Rag Picker's Daughter'—'tis here that the first step is taken which leads to infamy."[34] Still, owners continued to embellish the surroundings of their establishments in order to make the ice cream they offered more attractive. Joseph Campbell opened his ice cream saloon on July 4, 1855, and boasted that "his rooms are fitted up handsomely, his creams are richly flavored."[35] In the summer of 1856, George Barr bought the Drove Tavern, outside of Philadelphia, and "had it nicely fitted up" as an ice cream saloon. Lest his potential patrons be discouraged by the establishment's former function, his advertisement assured them that the neighborhood "is a good one."

Customer loyalty and regular patronage guaranteed the success of these businesses, explaining why owners used familiar advertising rhetoric year after year. To drum up business for his Maple Hill Retreat in 1858, Robert Loughead put on a "celebration" with cotillion and military music. "This beautiful situation, and the excellent quality of the ice cream furnished are not sufficiently known to be appreciated, or the house would be crowded daily with visitors," claimed the ad copy.[36] Yet providing patrons with pleasures could be a slippery slope. Like the pleasure gardens' lush bowers, which hid sexual liaisons, ice cream saloons, too, potentially fostered licentious behavior, an aspect also emphasized by popular novelists. In Ned Buntline's sensational "The Death-Mystery: A Crimson Tale of Life in New York" of 1861, the cheating wife of a gambler chooses "one of the most fashionable public saloons on Broadway: one where many a wrong has had its birth." During the hour of her "anxious" wait, she keeps ordering ice cream, "lest sitting so long alone should excite suspicion,"[37] making both the setting and the substance complicit in her sexual betrayal. Owners of ice cream saloons increasingly catered to a female clientele who did not patronize local taverns. Yet popular culture continued to express concern about these places of feminine refuge, because they allowed women some freedom. It is not surprising that institutions encouraging female autonomy were met with the same kinds of sensational discouragement as those promoting the autonomy of children in a culture that closely observed and tried to regulate pleasure in people's lives.

During the 1850s and 1860s, ice cream saloons funded lavish, summer opening celebrations reminiscent of those once offered by pleasure gardens, and in the North these continued even during the Civil War. Mrs. Balduff's Ice Cream Saloon, for example, hired the Union Band for her 1862 grand opening, using components of the war as a melodious backdrop for her confections.[38] The enduring success of northern ice cream saloons during the war also evinced on a greater level the North's success in the war itself and the degree to which it still enjoyed reliable supplies of fresh food. Local businesses, in fact, took their cues from the Lincolns, whose luxurious dinners provided inspiration for enterprising confectioners.

One such banquet, on February 20, 1861—just a few months before the onset of the Civil War—occurred at the Astor House in New York City. Lincoln and his wife enjoyed boiled salmon with anchovy sauce, terrine of goose liver, beef stuffed with peas, chicken with truffle sauce, duck, quail, beets, lettuce, celery, charlotte russe, French cream cakes, custard, and fresh fruit. Ice cream, the final dish, completed the feast.[39] Local papers like the *New York Herald* described the event in mouth-watering detail, down to the design of the menus and card stock they were printed on. Popular culture's preoccupations remained focused on refined tastes rather than politics.

A year later, Mrs. Lincoln hosted her first ball in the White House, described as a "snobbish imitation of the style of European Courts." The *Charleston Mercury* reprinted an editorial from a northern abolitionist paper decrying the first lady's "extravagant and foolish display" of luxury and extravagance. At midnight, the article recounted, she called her guests to the table and treated them to "one of the finest displays of gastronomic art ever witnessed in this country," which included a Temple of Liberty and a war steamer molded in candy, and a "ton" of turkey, ducks, and other poultry dishes. "At any time, such mimicking and aping of European courts is disgusting in the capitol of a republic; but at such a crisis as the present, such a wanton display of extravagance and indifference on the part of the administration is an outrage to the interests and feelings of the people," the paper commented. While Mrs. Lincoln and her guests were "merry with wine, jolly and indifferent," it observed, "that same night wounded volunteers died in the hospitals for want of care and comfort."[40] The lavish dinner provided the locus around which political issues were brought to light, symbolizing the many things people hated about the northern administration.

Frivolity in the South proved even more of an issue. In March 1862, the *Charleston Mercury* stressed "the limited supply of ice now in the Confederate States" and lobbied for its exclusive use by the military: "In a few weeks of the summer season the various hotels, ice cream saloons, and soda water fountains, would exhaust the present supply, and deprive the poor suffering fevered soldier of this most indispensable necessary. . . . [Ice] should be reserved for the use of the hospitals."[41] By contrast, not two months later the *New York Herald* hailed the imminent opening of a new opera house, which featured an ice cream saloon on the ground floor.[42]

Northern ice cream saloons continued to prosper, with new ones cropping up each season. After the battle of Gettysburg, one newspaper reported that out-of-town summer watering places in the North were doing a booming business, and that "an immense amount of money was expended by our shoddy aristocracy."[43] Balduff's Ice Cream Saloon, located near Philadelphia, flourished during the war, and by 1865, it had expanded to include a full confectionery, which supplied not only "Ice Creams of the choicest flavors," but "Candies, including Lemon, Hoarhound, Flaxseed, Walnut, Cocoanut, and other Candies and Taffeys."[44] After being fitted out "regardless of expense, with a sole view to the accommodation of the public," Morrison's Saloon opened in Chester, Pennsylvania, in the summer of 1864, catering to ladies, gentlemen, and "select parties" which "will find apartments especially provided for them."[45]

It remained business as usual for these fine ice cream saloons through the end of the 1860s and into the 1870s. They still carried out their operations on a fairly small scale, sometimes combined with candy and catering enterprises, and with only a few people responsible for the entire business. Confectioners produced their ice creams, cakes, and candies on site, in the basement of the same premises or in a separate smaller "house" attached to the main building. Some especially industrious confectioners supplied goods to the local hotels and catered private parties. They even facilitated the newfound mobility of the upper and middle classes by furnishing their outdoor excursions. In 1866, Mrs. Hall exhorted "Families, Parties, and Picnics, supplied with Ice Cream, Water Ices, Fruits, and other refreshments, at short notice."[46] Rice's Bakery and Confectionery, in 1868, provided "Parties and PicNics at wholesale rates with Confections, Ice Creams, Water Ices, and all seasonable edibles."[47]

The ability to accommodate a range of situations and a variety of customers kept successful confectioners in business. Enterprises like J. R. Munshower's in 1869 even supplied wedding parties "At The Shortest Notice."[48] As ice cream eating became democratized, the contexts of its consumption became more varied and required confectioners to adapt to a wide array of situations extending beyond the walls of their "recently done up" saloons. They now had to supply parties, picnics, hotels, balls, and private families with whatever goods their clientele desired. This meant not only providing the confections themselves but also delivering the accessories required to eat them: spoons, dishes, napkins, and even floral centerpieces.

Technological advances also aided the confectioner and caterer, yet the manufacture and distribution of ice cream remained a fairly localized affair through the century. In 1851, Jacob Fussell, of York, Pennsylvania, was the first one to apply steam power to the manufacture of ice cream. Twelve years later, he moved to New York City to capitalize on his innovations. Joseph Burke also ran a midcentury ice creamery at his Media, Pennsylvania, manufactory. Living outside of Philadelphia on a large farm, Burke had immediate access to ice cream's raw materials, including ice, fresh fruit, and new milk, which he bought from neighboring farms. He used a steam engine to turn the cranks of his many ice cream freezers, which enabled the cream to freeze smoothly within twenty minutes. During the 1868 season alone he sold 50,000 quarts of his ice cream to a Philadelphia confectioner, twelve miles away, who then sold it both wholesale and retail.[49]

Yet technology did not alter the social aspects of ice cream except to make its consumption more universal. Early in the century, the elite ate their ice creams in the lavish pleasure gardens. For most of the nineteenth century, the middle classes joined them, eating their ice cream at saloons that catered to specific comforts, desires, and incomes. Lower-class places, cropping up during the 1850s, consisted of storefronts, rooms in people's homes, and even street carts.

The Soda Fountain

In 1867, a Bostonian, G. D. Dows, published a picture of his award-winning "Appareil L'Eau Crème Glacée Soda," his Ice Cream Soda Fountain, in the *Druggists' Circular and Chemical Gazette.* Although it turns out that Dows's apparatus merely chilled the soda into which cream was added, his device and his name for it presaged the development of the ice cream soda less than ten years later.[50] Popular lore supposes that the ice cream soda as we know it first appeared at the Franklin Institute Exposition of 1874.[51] While this is debatable, it is certain that the ice cream soda fountain was already a popular institution by the mid 1870s, with the convergence of the soda fountain, the pleasure garden, and the ice cream saloon. The first soda fountains in America, in the 1810s, consisted of spigots mounted on the wall of the local apothecary's, or drugstore, and dispensed carbonated water alone. In the 1830s, druggists added flavorings and sweeteners to carbonated water, making it a more palatable beverage; like the penny candy, this nascent confection also began as a medicine. Uniting ice cream and soda water into one concoction created not only a new treat—the ice cream soda—but also a new institution—the ice cream soda fountain. Early pleasure gardens may have inspired the harnessing of nature in the form of public parks and later amusement parks, but as exclusive places of respite, they gave rise to the ice cream soda fountain. Like pleasure gardens, soda fountains also served as domesticated places of refuge, pleasure, and entertainment that were semi-public—open to all, yet catering to different crowds of people based primarily on gender and class.

Apothecaries and druggists first dispensed soda water as part of their therapeutic arsenal. Early on, these soda water dispensaries constituted hubs where important people of the town gathered to discuss political and social matters; they were almost exclusively male bastions. When apothecaries added flavored syrups to their sparking waters, carbonated beverages became popular with a broader audience, especially women, who enjoyed them as much for their sweet taste as for their purported restorative properties. When Robert Green, a soft drink concessionaire, added ice cream to soda water at the Franklin Institute Exposition later in the century, his ice cream soda became an immediately popular treat, and the soda fountain grew from a spigot at the end of a druggist's counter into an institution that replaced the ice cream saloon.

The soda fountain provided a clean, lavish haven for those who could afford a 15-cent egg cream or 10-cent water ice. Soda fountains became key meeting and socializing places for the urban elite, and they also frequently sold candy, bonbons, and tobacco. Many soda fountains, located in department stores, offered refreshing libations to the tired female shopper. Others carried an assortment of tobacco products and allowed only men. Some defined themselves as temperance establishments when that movement gained momentum, and many of those places targeted the taste buds of men either by promising them healthful alternatives to alcoholic drinks or by covertly serving them alcohol disguised as or in soda water.

Fashionable, highly social spaces that served "dainty" foods, soda fountains continued to reinforce the cultural hierarchies among consumers that had been established well before the beginning of the nineteenth century. The way confectioners decorated them to appear as urban oases enabled their clientele to escape the sordid elements of the city while in the city. These spaces presented new attractions—no longer in the form of nightly illuminations or cultivated plants—but as technology itself. Soda fountains, which incorporated gas lamps and running water, were ultradomesticated, artificial environments that celebrated the technological ideal of man's triumph over nature. Like the pleasure gardens and ice cream saloons, soda fountains sold their atmosphere as much as the food itself, and offered treats for both the eyes and the palate. What is more, the ambiance of these establishments became increasingly distinct as the century neared its end, so that "men's" fountains resembled traditional taverns, while "women's" fountains, much more prevalent, incorporated the display ethic of the department store.

The success of "women's" fountains only helped to reinforce popular associations of women with sweets and gave them comfortable places in which to consume them. Removed from the businesslike context of the local drugstore, the soda fountain became a popular attraction for fashionable women in urban America. For refined women, the soda fountain operated as a place

FIGURE 15. Trappings of the turn-of-the-century soda fountain included state-of-the-art dispensers, immaculate surfaces, and lush plant life. "Soda Fountain, Centralia, IL," *Soda Fountain,* June 1907, 21. Courtesy of Harold and Joyce Screen.

of rest and socializing during a day in the city. Tailored to the genteel feminine aesthetic, its appointments included potted plants, clean counters, and attractive soda jerks (fig. 15).

Fountain owners and their female customers shared a complex relationship. While most operators did well if they set up shop in busy commercial districts to attract the many women bustling about the streets, they also relied on a quick turnover to maximize profits in the summer months, their peak season of operation: "It goes without saying that the major part of the income to the soda water dispenser is the income that is either directly that of womankind, or that is influenced by her."[52] Women could be counted on to find daily refreshment at the local fountain. Many fountain operators employed lavish decorations, state-of-the-art fountains, and marble-clad interiors in order to draw their crowds and justify the 10 or 15 cents they charged for an ice cream soda. Trade journals for fountain operators suggested various ways an owner could attract and keep his female customers, who, they agreed, could be fickle.

Professionals considered the very design of the mechanical fountain apparatus itself to be important in garnering this trade. In many instances, the more popular designs—constructed of different colors of marble and featuring water jets, figurines, and ornamental bric-a-brac—proved less practical and durable than their simpler counterparts. But in the 1870s, ornate styles reigned supreme. Businessmen saw them as solid investments that drew the crowds and hence the profits. An article published in 1878 detailed the "Inferiority of Applied Ornament" on fountains, because it was both fragile and difficult to clean. The author wrote, "Beyond a certain moderate limit, the money paid for a 'soda' water dispensing apparatus is paid solely for style—for that which has no value except as it strikes the public eye agreeably."[53] The "good" style, more Spartan, contradicted the "bad" style embellished with detailed decorative motifs (fig. 16). Although the "good" style preceded the streamlined fountains popular twenty-five years later—characterized by plain marble slabs and simplified design elements to emphasize cleanliness—the "bad" style still remained the more popular model among marketers of soda fountain apparatus for over twenty years, because it contained all the interesting details and excessive ornamentation that made the fountain itself an alluring visual attraction.

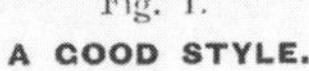

FIGURE 16. Changing fashions were not just confined to women's clothing. "Soda Water Styles," *Carbonated Drinks,* January 1878, 30. Library Company of Philadelphia.

Many fountain devices incorporated nude female figures, glass domes with spurting jets of water, and gas lights. Fountain manufacturers took advantage of people's fascination with the triumph of technology that miraculously delivered fizzy water to beverage glasses by offering other gadgets that used water in novel ways. For example, the countertop Revolving Tumbler Washer and the Crystal Spa, both illustrated in Charles Lippincott's 1876 trade catalogue, washed and rinsed dirty tumblers right in front of the customers. The drudgery of cleaning dirty dishes, usually hidden in the private back room, became a public attraction performed in plain view. The Crystal Spa, topped with a gas light, cleaned tumblers "inside and out." The ad copy continued, "its novelty and beauty must be seen to be fully appreciated . . . the value of an article like the Crystal Spa can only be justly realized by regarding it as a thing of beauty merely, apart from its mechanical character." Like the Crystal Spa, the Revolving Tumbler Washer operated on the same principles, doing "its work thoroughly as a washer, and at the same time its activity gives a pleasing variety to the counter arrangements."[54] These apparatuses distilled the most ungainly aspects of mechanization—the valves, belts, tubes, and even the huge tanks of compressed carbonic acid and water, necessarily housed in the basement—into entertaining spectacle. These kinds of feats made technology an interesting and accessible attraction for women, in particular, and brought the wonders of industrialization that they had witnessed at events like the Centennial Exposition into closer view.

Fountain apparatus itself provided visual interest that recalled the beauty of nature but tamed it, using natural characteristics as advertising points. James Tufts's "Arctic" soda water models emphasized coolness in the heat. "The Alpine," for example, stylistically recalled the steep snowy slopes of the Alps, with its light veined marble and iciclelike hardware. Competing with Tufts, Charles Lippincott's devices suggested the lush green worlds of the spas; some of his fountains came embellished with pictures of tropical scenery. "The Pandora," for one, displayed a mural of etched mirror, replete with herons and palm trees. One's gaze into the surface reflected back to one amid the romantic lushness of a far-away place.

By the late 1880s, people no longer wanted fountain apparatus that referred to nature. They preferred machines that celebrated technology for its own sake. These later fountains—simpler, less ornamental, and more in line with the increasing popularity of the simplified modern aesthetic—valued spare-ness over embellishment as much to change the definition of what was "fashionable" as to emphasize and demonstrate cleanliness and sterility. Sleeker expanses of smooth marble and bold spigots replaced gurgling waterfalls and nude statuettes.

Businessmen remained conscious of the preferences of their female patrons, who, as active shoppers, made it their business to keep abreast of the latest styles; it is not surprising that fountain apparatuses, too, followed the fashions of the time. In fact, many people directly connected women's fashion with the styles of the soda fountain, further reinforcing the woman/confectionery connection. One manual offered this advice about the soda fountain's appearance, "'Costly though thy apparel as thy purse can buy' can be well applied to the garb which covers and preserves that delicious beverage."[55] Another analogized that an ornamented fountain resembled a vulgar woman, "tricked out with a profusion of useless bracelets, necklaces and finger rings," whereas a classier fountain, whose details "are made as graceful and attractive as possible" was "clearly in better style," like the woman whose "jewelry takes the form of a gold watch and chain."[56] Not only should the soda fountain resemble the woman in appearance, but it also should reflect her impeccable taste. Producer and consumer shared consonance: if, the comparison went, one wanted to attract a refined patronage, one needed to match the aesthetic of the apparatus with the sensibility of the customer. Therefore, fountains both valued and resembled the woman wearing gold, diamonds, and pearls, rather than those "tricked out" with cheap imitations.

Many soda fountain operators relied heavily on the interior decoration of their establishments to attract the female trade, providing a clean and comfortable atmosphere into which women could feel confident about bringing their children as well. One man described this succinctly as the need to provide "tasty surroundings."[57] Owners of the most popular establishments outfitted them with luxurious plants and fine marble counters, because appearance and taste equally stimulated trade. One reputable advisor emphasized using plants for decoration, perhaps because they suggested the old oasislike pleasure gardens. Businessmen felt that appearance equaled taste as important factors stimulating trade: "Plants, such as ferns, potted palms, rubber plants, etc., should be secured and placed about the apparatus. The sight, as well as the palate, shall be pleased, and thus the beverage made to appear more delicious. The plants assist in imparting a sense of refinement and fragrance to the establishment."[58] In contrast, owners seldom took into consideration men's taste preferences, except to confine them within the dark, smoke-filled, wood-paneled, and trophy-laden walls of the tavern. Soda fountains remained female bastions.

Interior decoration alone did not attract the female trade, so many owners relied on sex appeal to procure the business of women. By "outfitting" their establishments with young and attractive servers, fountain operators secured not only the patronage of married women who were continuing their window shopping inside but also attracted younger female workers who were between work shifts. Many business manuals and trade articles advised owners to find help that was not only reliable but physically appealing. R. R. Shuman, writing on the "Psychology of the Soda Fountain," asked, "Is it the 'perfectly stunning' boys behind the fountain?" that drew the female trade. Another manual suggested that a soda fountain attendant "should preferably be one comparatively young in years and prepossessing in appearance." While the fountain operator should "be always friendly," he should not, however, "attempt familiarities with his patrons."[59] Still another stated, overtly, "Place a handsome young man behind the counter. He'll draw the best trade in a city if he strictly minds his own business and dresses neatly and fashionably."[60] In order to encourage repeat visits, the attractive counter boys would presumably draw women who came again and again to fawn; however, if they showed allegiance to any one of them, they risked losing their business altogether.

Appearances in the fountain also assured women that the goods served to them were pure and their surroundings were clean. Sparkling glasses, gleaming silverware, streakless mirrors, and greaseless marble countertops made this concept physically manifest. By the 1890s, when issues of sanitation reflected broader concerns of late-century reformers, advisors laid heavy emphasis on the importance of fountain cleanliness. The cleaner the fountain, the less one risked contracting an illness due to dirty silverware or dishes. In addition, an owner who kept his fountain sparkling seemed less likely to use artificial (and potentially poisonous) flavors and colors, and adulteration remained a chief concern not only of pure food advocates but also of those leading the small yet unceasing backlash against luxurious indulgences, including confectionery (fig. 17). "Soda water is a matter of taste,—in a double sense. Scores of customers lose their thirst when they see a slovenly fountain. Keep your fountain clean," warned one manual.[61]

"Dangerous habit-forming drugs are sold freely to women and children."

FIGURE 17. Moral and physical corruption always lurked at the soda fountain. "Death Serves Soda," *Soda Fountain,* June 1909, 27. Library of Congress.

Prescriptive literature aimed at late-century soda fountain operators continually emphasized eye appeal as a way to communicate more abstract ideas to women, like cleanliness. "Daintiness" served as a common euphemism for cleanliness. "Thin, sparkling glasses, bright, shining silver holders, new daintily designed spoons, clean tables, spotless jackets, immaculate serving counter"— all contributed to a successful fountain. Yet the clear message was this: "You may talk to a woman until you are blind about Pure Food beverages, about chemically pure carbonic acid gas, about ice cream made from rich cows' cream and nothing else, and you may give her all these, but if you do not serve them daintily—you lose."[62] Proprietors interested in attracting a more sophisticated trade appealed not to woman's reasoning but to her discerning eye instead. The message to a woman about healthfulness was not conveyed through scientific appeals, because she would not be able to understand them. Appearances, more than talk or text, convinced a woman of the safety of her environment.

All fountain operators aspired to attract a genteel trade who could afford to buy more expensive ice cream sodas, who appreciated the social atmosphere, and who also might purchase corollary products like bonbons or sandwiches. Time and again, owners depicted their businesses as places of retreat from the outside elements, which included the sweltering heat of the urban summer. Coolness subsequently joined with cleanliness as a highly valued feature contributing to a fountain's success: "To keep the store cool during the days of the heated term is, or at any rate ought to be, the aim of every progressive and up-to-date retail merchant," suggested one turn-of-the-century article extolling electric fans.[63] By the beginning of the twentieth century, many fountains equipped themselves with electric fans to keep breezes blowing gently across their feverish customers. "The cool air may be preferred even to the cold drink, especially when the price of the latter pays for both," wrote tradesman A. Emil Hiss.[64] Operators also employed other techniques to convey the idea of coolness, including outfitting the interior with marble and mirrors, and serving confections in freshly polished silver- or chrome-plated dishes and glass holders. Coolness became such an important factor in people's lives that they developed an expertise for discerning different degrees of coldness in their drinks. Many would order sodas with various amounts of ice to achieve a confection chilled to their individual taste.

All of these elements helped attract a customer base made up chiefly of affluent and semi-affluent single women and mothers. But these alluring features could also be drawbacks to the very proprietors who lured these groups. On the most successful (and hottest) days, the larger fountains reported selling upward of 1,000 glasses a day, and they needed a rapid customer turnover in order to see viable profits from the pennies garnered from each transaction. Therefore, the very visual attractions they used to entice women became deterrents to the brisk business required on hot days. The ice cream soda, the most expensive treat at the fountain, also required the longest time to consume. Ordinarily, it was a profitable confection that encouraged additional purchases and repeat business. But on busy days, the woman consuming an ice cream soda could be the bane of the businessman: "Time is everything to a soda man on a hot day. . . . It takes a woman considerable time to eat and drink her glass of ice cream soda. Usually she expects to make this an excuse to rest and gossip, and she may occupy her seat at the table or counter for ten or fifteen minutes. . . . It is convenient to be out of ice cream when the mercury is climbing up among the nineties."[65] Being "conveniently out of ice cream" meant that a proprietor could increase his turnover by serving plain sodas and thus maximize his sales without offending the woman who wanted to linger over her ice cream soda.

But wise fountain operators continued to be sensitive to the desires of their female patrons, who comprised the bulk of their business. Students of the Monticello Girls' College in Illinois, for example, boycotted the local soda fountain in 1908 because the owner had raised the price of an ice cream soda to 10 cents at the beginning of their fall term. "The girls decided they would wait until they went to St. Louis, not far away, for the week end visits, and would drink enough there to tide them over the week." When the students returned after the Christmas holidays, they found their resolve stronger than the shop owner's, and "were greeted with large signs announcing 5 cent soda."[66] Such was the increasing presence and importance of women as consumers, and proprietors' increasing willingness to cater to them.

Women consistently found a congenial atmosphere at soda fountains installed right in department stores. The stands numbered among the several amenities that store managers provided for their predominantly female clientele, which also included spacious lounges, lavatories, and restaurants "catering to a woman's palate. These were places where shopping women could meet and chat with their friends, comparing purchases and trading tidbits about the latest styles."[67] Not only did soda fountains provide a relaxing and profitable place for women to rest from their activities, but they also kept women from leaving the store at all. Department store fountains, marble-topped islands within the larger retail environment, perhaps most successfully identified and targeted middle-class female customers and capitalized on their impetus to shop and socialize at the same time.

Selling ice cream sodas to a regular male clientele, however, was less easily accomplished. Although the temperance movement discouraged tavern business, neighborhood saloons selling alcohol were still the soda fountain's main competitors when it came to male customers. In addition, by the end of the nineteenth century, most fountains and taverns had become distinctly gendered spaces in the minds of patrons. It may have been permissible, or even acceptable, to eat ice cream while reclining on velvet couches with other "dandys" at Parkinson's Saloon in the 1850s and 1860s. But by the 1880s and 1890s, this was no longer the case, as Gilded Age Americans increasingly segregated themselves based on a number of demographic factors, including age, class, gender, and ethnicity. These differences appeared in and affected all aspects of late nineteenth-century life, from determining domestic roles and type of employment to the very things that people purchased and what commercial environments they frequented.

Trade journals dedicated to soda fountain operations at the turn of the century explained how to capitalize on new market territory by targeting male customers who had previously spent their time in taverns. Some taverns, desperate not to lose business to the temperance campaigns, installed soda dispensers in their establishments and served sodas alongside alcohol. While this may have instilled the idea that soda water concoctions were pleasurable to drink, it also inspired a new breed of mixed drinks that combined alcohol and soda water, undermining reform efforts. The temperance movement helped soda fountain sales by suggesting alternatives to alcoholic beverages. One writer optimistically saw this as a new era in which the sale of soda water would greatly increase, and that taverns might actually serve ice cream sodas as well.[68] Paradoxically, however, soda fountains both offered a genuine respite from the dangers of drink and also, in some cases, became covert dealers in alcohol themselves.

In most cases, soda fountains remained institutionally separate from alcohol-purveying saloons, and savvy entrepreneurs tried to attract a new consumer base by founding establishments tailored to the masculine aesthetic. Temperance workers recognized how beneficial soda fountains could be to men, but conceded that they had to be divested of their distinctiveness as feminine preserves. Ironically, while reformers railed against alcoholic-inducing penny candies for children, they believed that sweetened soda water delivered men from their liquor habits. Suddenly, sugar-laden sodas were not addictive and, according to temperance advocates, discouraged a taste for alcohol altogether. Writing rosily about the fountains' possibly changing clientele, the reformer Frank Harmon wrote: "The associations, too, are different, and instead of being patronized solely by women and children, it is the men who are now turning to the soda fountain, with its refreshment that leaves the busy merchant, the banker, the broker, invigorated and clear-minded for the rest of the day's business." Other reformers saw soda fountains not only as an alternative to the saloon but also as active weapons against alcohol consumption: "At the soda fountain, one or two drinks suffice. In the saloon, it becomes a question of money, credit, or physical ability, to the degradation of the victim's moral, mental and physical welfare. Every soda fountain installed is one more working element in our campaign."[69] This was, of course, the optimistic rhetoric of the temperance cause; many soda fountains continued to offer beverages with a kick. Philadelphia police even arrested some fountain operators for selling liquor illegally.

Through the first decade of the twentieth century, trade publications continued to record soda water's victory over alcohol. For example, the *Soda Fountain* happily reported in 1908 that Detroit had been very successful in enforcing "dry" laws, because the police had switched from targeting bars operating illegally on Sunday to soda fountains covertly selling "cocktails, highballs, gin fizzes and the rest." The report continued, "At some Detroit fountains, in the past, it has been possible to secure straight whiskey, it is said, but owing to the present state of the public mind this is no longer true to any great extent."[70] But getting men into "dry" soda fountains proved a hard sell, especially since the fountains had been so successful in establishing themselves as feminine retreats, fit for women and children enjoying their leisure time.

An enterprising pair in New Orleans decided to combine the soda fountain and tobacco store, creating a clientele that was about 95 percent male. The modest shop dedicated little space to a seating area. The proprietors said that they enjoyed the best of both worlds because they catered to a male population but could also accommodate women's tastes: "A man who feels a delicacy in leaving a lady to go into a tobacco store, brings the lady with him into our store. The result is a sale of tobacco or cigars and a sale of soda water or ice cream. It has also proved convenient for men after eating ice cream or drinking chocolate, to step to the cigar counter and get a smoke. We have found that the two go well together."[71] In order to increase their female trade, the owners had "done things we know pleases them. Everything is strictly orderly and scrupulously clean."[72] What probably caused most of their success was the

FIGURE 18. Soda fountains for men adopted the tavern aesthetic and often sold tobacco as well. "Combination Tobacco-Soda Fountain," *Soda Fountain,* February 1909, 31. Library of Congress.

reluctance of men to imbibe soda waters or ice cream sodas in a more feminine setting. By offering "feminized" confections such as ice cream sodas alongside "masculine" products such as tobacco, and within a space whose gentility had been deemphasized, the New Orleans businessmen created a comfortable atmosphere in which men could indulge their collective sweet tooth. Contemporary photographs of these "male" fountains show them decorated like regular saloons (fig. 18), with heavy trim, dark wood paneling, and stuffed animal trophies, unlike the mirrored murals, flourishing plants, and sparkling marble of "female" fountains.

A Chicago fountain operator remarked on the increased sales of fountain drinks to men after the turn of the century in his city. He enumerated the reasons why men patronizing soda fountains were no longer a "rarity," and asserted that "temperance has got hold of the Chicago man." He also claimed that it was cheaper for a man to drink soda water than "something stronger."

Finally, he pointed to changing social habits, because "more men go shopping with their wives now than formerly. . . . Most men can't stand tea, and a compromise is made on ice cream soda for the women and a sparkling drink of some kind for the man."[73]

But the fountains had become so closely associated with women that people still considered a man's imbibing there to be an aberration. A 1909 trade journal article tried to debunk these presumptions about gender by arguing that more men than women were now eating ice cream sodas and sundaes at one Milwaukee fountain. One of the proprietors contended that "mere man absorbs three-fourths of the sweet stuffs." The reporter continued, "In many of the drugstores of Milwaukee the soda trade seems to be primarily supplied by men. It is they who possess a sweet tooth and the capacity."[74] It is impossible to say who the key consumers of these confections were, because there are no contemporary consumption statistics based on gender, but by the turn of the century, American culture clearly assumed the primacy of women in this regard. Even if they were not, sweets had become so inextricably linked with females that it seemed natural to assume that women consumed ice cream sodas and men used tobacco.

Age-related factors also helped define the viable fountain patron, including courting teenagers, further cementing the connection between sweetness and pleasure. As early as the 1830s, the romantic intentions of the young created part of the dynamic of the emerging ice cream saloon. One Philadelphia chronicler described a local fruit stand and candy store during the 1830s, "at which they sold ice cream in summer and to which the very boys of the neighborhood would take their best girls and calling for one cents worth of cream and two spoons, the girl with her spoon sat on one side of the table and the boy on the other. Of course the girl got the biggest share of the cream."[75] Another teenager later recorded in her diary that she cut choir practice and sneaked out to the local soda fountain to drink "cream soda and have an elegant time" with the boys.[76] Ice cream could be shared in public and provided the reason for innocent romantic encounters. Between a boy and a girl, it was a treat less fraught with intimate romantic symbolism than a box of chocolates. Yet the atmosphere of the ice cream saloon and the later soda fountain remained openly sensual, emphasizing texture, temperature, sight, smell, and taste—elements not valued in regular taverns of the time.

Some fountains not only transcended age barriers but class barriers as well.

By the turn of the century, the local soda fountain no longer served the comfortable classes exclusively, bringing cool treats to the economically depressed as well. One example was a fountain operating in a "Chicago Ghetto" after the turn of the century whose interior appeared clean but spare—a different world in class and aesthetics from the upscale fountains in shopping districts and department stores. The proprietors of this Chicago saloon offered less expensive confections to a wider population, "catering to all classes of people who enter the store." Their ice cream sodas and sundaes cost a nickel less than those sold at upper-class businesses. The proprietors acknowledged that "there is a very small per cent. of profit in them, but if the sales are large enough they pay." The fountain's business responded to the work patterns of its lower-class neighborhood, fluctuating during employment cycles and work hours, selling most successfully in the summer and on Sundays. "The recent hard times affected my business to a certain extent," one of the proprietors said. "The reason for it was that the people in this locality are mostly laborers." In defense of his clientele, he added, "The laboring men out here spend more money at the fountain than they do in the saloons, contrary to the prevailing opinion of those who do not know."[77] By the beginning of the twentieth century, soda fountains were no longer the exclusive domain of the elite. The soda man could "tell you of some interesting facts about the delicate hands of the rich ladies of the vicinity and the rough hands of the washerwomen who come to the same fountain. . . . There are soiled hands at the fountain, and there are hands which have been manicured to the extreme."[78]

Ice Cream Vendors

When it came to eating in saloons and fountains, people paid as much for the atmosphere as they did for the confections: the context largely determined whether a dish of ice cream cost a penny or a dime. Street vendors, popular since the late 1830s, charged much less for ice cream than establishments with mirrored walls and marble-topped tables. Just as the ice cream saloon and soda fountain became familiar institutions among the fashionable, street vendors became familiar among laborers. One man noted in 1847 that in addition to the numerous "Creameries" appearing in just about "every other house," he found ice cream outside as well. "The stalls of the markets are crowded with it—it is wagoned about the streets, and turn where you will, there is ice cream in profusion—cheap but not the less excellent on that account."[79] Unlike saloon owners, who relied on the word of mouth among their refined customers to generate business, the street vendor publicized his offerings by shouting them to the world. That this practice often annoyed urban dwellers highlights the differences between public and private selling and suggests the disparities between public and private consumption. For example, the author of *City Cries,* published in 1850, criticized ice cream vendors for being too noisy: "These gentlemen, however, although they contribute to the gratification of the public by their excellent confectionery, do not contribute to their amusement, by crying their good things in the streets."[80]

Yet nineteenth-century urban culture continued to hear the calls of the street vendor, as he catered to a much different segment of the population. The pushcart vendor was not only poor himself but served the poor, often members of the immigrant community to which he belonged. Significantly, street vendors threatened institutional order by providing alternative markets to the enclosed shops of the saloons, fountains, and department stores by creating public gathering places for the urban masses. An 1868 illustration in *Harper's Weekly* depicts the class disparities determining and determined by consumption contexts, juxtaposing a genteel interior on Broadway with immigrants in the Bowery Garten (fig. 19).

Politicians and reformers railed against the street vendors, accusing them of causing urban congestion and selling unsanitary products. Yet they were really concerned with the vendors' potential disruption of a social order that maintained class-based hierarchies. "Pushcart markets provided immigrants with a free social space unburdened by the financial costs paid by wealthier city people both for larger private residences and for semipublic places constructed for social gathering and amusement," one historian observes.[81] Maintaining the boundaries of these "social spaces" enabled the economically comfortable to retreat into their own worlds—pleasure gardens, ice cream saloons, and soda fountains. Operating in public, pushcart vendors threatened the status quo by blurring the lines of property and class, reminding those who sought refuge in private places why they were retreating to begin with.

Ultimately, the ice cream vendor, like many other itinerant sellers, encroached upon the financial, physical, and psychological territory of the elite. More than other noise pollution, the vendor's street cries assailed upper-class sensibilities. A man writing in an 1855 issue of the *Sunday Dispatch* complained

FIGURE 19. "Up Among the Nineties," *Harper's Weekly,* 15 August 1868, 520. Library Company of Philadelphia.

about the "decided annoyance" due to the "vociferousness of the dispenser of frigorific delicacies": "I am annoyed nightly by an itinerant ice cream vendor crying his 'confections' through the street in which I reside. My business requires me to retire early, (nine o'clock) and at ten I am regularly awakened by his screams as he passes under my window, both going and returning."[82]

Pure food and drug advocates accused ice cream vendors of selling poisonous ice cream. True, ice cream did get contaminated with verdigris when tin-plated copper molds and canisters became corroded enough to expose the underlying copper, which reacted chemically with the acids in the ice cream, but this happened as often to the fine confectioner's product when he was not careful about the state of his molds as it did with the street vendor's goods. However, most publicized accounts of verdigris poisoning sensationalized the role of the street vendor and downplayed that of the established confectioner.

Although the "hokey pokey" man was a popular site with the local children, who could buy their ice cream in penny dishes, he was a blight to the reformers, who believed him to be the chief source of ice cream poisoning.

They repeatedly accused him of refreezing leftover ice cream that had liquefied to sell the next day. "No matter how good when made," one man wrote in 1900, "decomposition will set in when it liquefies, and ptomaines are likely to be generated and other injurious effects will occur. Some makers re-ice old cream and sell it as 'hoky-poky' or 'Greco-Roman cream.'"[83] Some people, though, could only afford inferior goods. "Since hokey pokey was neither made nor sold under sanitary conditions, vendors became adept at dodging the public health inspectors."[84]

Charges against itinerant vendors also assumed a xenophobic tenor, making both the act of vending and immigrant status key rhetorical issues. Italian immigrants who came to America during the 1880s brought a culinary history that made them particularly proficient at manufacturing ices and ice creams,[85] which they sold on the streets. But doing so exposed them to the wrath of late-century reformers who used the real and imagined characteristics of immigrant communities as scapegoats for prevalent social problems. After recounting two cases of ice cream poisoning in Europe and agonizing over the "thriving business among school children and street vendors," a writer in the summer of 1902 assailed the "habitual filthiness" of Italian immigrant ice cream sellers in the United States, saying:

> Few people realize what a menace to health is found in the sale of this delicacy. The commonest and stalest materials are used in its manufacture, and at night it is usually stored under the merchant's bed in his dirty tenement lodgings. The next morning, no matter how far gone in decomposition the unsold cream may be, it is rehashed and frozen for the day's business. . . . In addition to all this, every opportunity is afforded for transference of diseases between the customers, for the glasses and spoons are never washed, but are merely rinsed in water that accumulates the filth of the entire day.[86]

Refined people paid extra for their ice cream precisely because they did not want nor have to associate in any way with the "filth" of the Italian vendors or the sociopolitical implications of their "dirty tenement lodgings." Ice cream's additional cost in fancy establishments paid for reassurance: the ice cream was pure by association with its pristine surroundings—cleanliness made possible by Anglo-American innovations.

Italian vendors posed a threat to children through yet another product. In the first years of the twentieth century, responding to the very health concerns that reformers raised, many vendors took to selling ice cream sandwiches made by layering a slab of vanilla ice cream between two wafers, using an ingenious presslike gadget. "The cream is not brought into contact with the dispenser's person in any way," the public was assured.[87] The *Confectioners' and Bakers' Gazette* decried the sight of the "gaudily painted barrow" of the Italian street vendor, who "knows the American youth, and knows they can be depended upon to put down their cents, and gobble up ice cream sandwiches with gusto." After describing the creation of the sandwich, the writer remarked sardonically, "Then in exchange for a cent, the product is handed over, and the transaction is complete—so far as the seller is concerned. What happens to the purchaser is another matter,"[88] leaving the reader with the impression that a dirty, poor, greedy Italian was willing to jeopardize the health of innocent children to make a profit. In reality, the vendor's business threatened the established and "reputable" confectioners who subscribed to trade journals like the *Confectioners' and Bakers' Gazette.*

Finally, under pressure from the ice cream industry, city governments went after street vendors. "The authorities of Philadelphia have instituted a crusade against the ice cream *peddlers* who trundle carts through busy Market and Chestnut streets, to the detriment of traffic," the *Soda Fountain* reported in 1906. "Many of them have been arrested, but were released when they promised to avoid those thoroughfares in the future."[89] Yet another account claimed that ice cream and pretzel vendors who congregated near public schools would "imperil the lives of the children" by selling them adulterated goods. "Principals and teachers will be granted permission to put up stalls and supply pure ice cream, pretzels and other things to eat to the children," the report continued.[90]

The larger manufacturers hoped to drive the smaller vendors out of business in order to ease competition. Since smaller vendors could realize only marginal profits, even by using cheaper substitute ingredients (which often were not harmful), established manufacturers coopted the rhetoric of the pure food movement to muscle competitors out, mandating the use of "pure"—meaning more expensive—ingredients.[91] One commentator wondered why the public attacked the vendors more than the "upstanding" businessmen of the city, who were as guilty of selling adulterated products as the poorer people: "Dwarfed Italians, who sell poisonous soda syrups, without knowledge, probably, of what they are doing, find their names emblazoned in print when warrants of arrests are issued for them; but the men who have the price of advertising in their clothes escape the punishment of publicity."[92] A 1909 trade journal reported that Pennsylvania's state dairy and pure food commissioner had "accused foreign makers and vendors of ice cream of producing the bulk of the inferior article and putting it upon the market." The commissioner then congratulated the Pennsylvania Association of Ice Cream Manufacturers for successfully "driving out this class of manufactures" and "complimented the organization upon the integrity of its own members."[93] Xenophobia, class distinctions, and marketing concerns all qualified as reasons why various interests spoke out against the itinerant street peddler, who lacked affiliation with professional trade associations and a stable business in a storefront.

While some people in all businesses sold adulterated goods or those made out of cheaper materials, not all street vendors sold tainted ice cream. Tacit in the critics' complaints was their objection to ice cream's ubiquity. Stripped of all attendant frills, it was accessible to all and no longer marked class and status. The *International Confectioner* more honestly identified social pretensions when it wrote in 1908 of publicly vended ice cream, "There are a great many who refuse to eat ice cream in this undignified manner but had they a plate or even a paper plate, a spoon and a little napkin, they would be tempted to buy ice cream from these places."[94] The customers' "dignity," according to this writer, stemmed not from *what* they ate but *how* they ate it. If the people who cared about image and affectation could eat ice cream with more genteel utensils, they would deign to purchase it from the conveniently located vendors they so cavalierly and vehemently decried. Debates about ice cream were really about ideas and ideals of respectability and cultural hierarchy. The elite found blurring these distinctions to be unacceptable because they had constructed an entire system of etiquette and social discourse to maintain cultural differences.

Street vendors, soda fountains, ice cream saloons, and pleasure gardens were the main institutions that brought ice cream to nineteenth-century Americans. But ice cream had also been made in elite homes since the end of the eighteenth century, and recipes for ice cream commonly appeared in nineteenth-century recipe books. Homemade ice cream was made more feasible by the introduction of hand-cranked ice cream freezers in the 1840s. Their invention, incorporation into domestic rituals, and impact on Americans' eating habits are discussed in Chapter 6, as part of the exploration of the uses of sugar in the home.

Other kinds of confections that were equally distinct repositories of sugar joined the democratization process along with penny candies and ice cream. Chocolate, the next confection to enter popular tastes during the nineteenth century, was similarly a form of confectionery both celebrated and reviled. While ice cream made available in the marketplace threatened comfortable class hierarchies, chocolate threatened to destabilize established gender divisions.

CHAPTER FOUR

Sinfully Sweet

Chocolates and Bonbons

In 1836, the partners Henrion and Chauveau of Philadelphia numbered among the first confectioners to sell fancy chocolates. They expanded their lines of confectionery beyond imported sweetmeats and comfits, giving their discriminating clientele the latest fashionable treats. Henrion and Chauveau did not cater to the developing and reliable sweet tooth of children by selling cheap penny candies; their business succeeded until the 1850s because they focused on a very different group of consumers by giving them a very different type of product—soft candies, chocolates, and bonbons. The Philadelphia elite could afford the luxuries men like Henrion and Chauveau offered, and they were a customer base worlds apart from that found at the neighborhood candy shop. Indeed, modern chocolate, the auspicious marriage of the derivatives of sugarcane and the cacao bean, was the most coveted confection among nineteenth-century Americans.

More than fifty years after Henrion and Chauveau opened their shop, Marion Harland, a popular and prolific domestic advisor who also lived in Philadelphia, understood that by then the taste culture had shifted. Chocolates and bonbons, no longer the exclusive treats of the rich, even reached the palates of teenaged girls. In 1889, Harland published a household advice manual that included recipes and general domestic hints based on her years of experience. Pointing out how soft candies affected both those who made and those who ate them, she wrote, "It is not an uncommon thing for a couple of school-girls to eat a pound of Huyler's 'butter-cups' . . . at one sitting. I have seen the belle of a summer resort dispose with apparent comfort of five pound boxes in as many days. So well is this passion of the maiden's soul understood by him whose life-long business it is to make her happy, that he feeds it with the regularity of grist to a mill, her ruby mouth being the hopper."[1]

School girls eating bonbons with brand names; resort "belles" consuming a pound of sweets a day; businessmen who understood a woman's "passion" and would feed her "ruby mouth": these numbered among the meanings that soft candies held by the end of the nineteenth century. Yet chocolate's meanings developed from centuries' worth of combined symbolism and sentiment from many different cultures. And like sugar, chocolate did not begin its life as a substance people considered feminine.

∴ *Chocolate's Dark History* ∵

Indicative of the high esteem in which chocolate was held, Carolus Linnaeus gave the Central American tree from which the cacao bean comes the genus name *Theobroma,* or "food of the gods." Chocolate's history, like that of sugarcane, is long and murky. It probably originated near the Mexican gulf coast around 1500 B.C. The Mayans, who flourished from A.D. 250 to 800, probably incorporated chocolate into their diets before anyone else. They ground the cacao beans, mixed the powder with water, and drank the concoction unsweetened.[2] During this same period, people in far-flung parts of the world were experimenting with modern confectionery's other raw materials as well. The Persians and Chinese were using ice, Hindus were making sugar from cane, and the Arabs were expanding sugar-making technology into Arabia, Spain, Morocco, and various African countries.

From 1486 to 1502, the Aztecs were busy conquering the province of Xoconocho, known for its capacity for cacao production. They, like the Europeans who soon followed, took advantage of the favorable climatic conditions in order to grow cacao, which they used both as currency and to make luxury drinks. Columbus, who had already brought sugar to the New World on his second voyage in 1493, stumbled upon cacao beans in Mayan trading canoes after having landed near Honduras in 1502 and quickly realized their value. Spain conquered the Yucatan and Mexico by 1519, in part to "take advantage of the monetary value of cacao beans in the native economy."[3] Eventually, like coffee and tea, chocolate became a fashionable beverage for elite Europeans. On account of its alleged properties as an aphrodisiac, a claim it has never completely shed, cacao held even more fascination for the Spaniards than did sugar.[4] Regardless of purpose—money, sex, gustatory pleasure—people ingested chocolate primarily in liquid form through the nineteenth century. And until the mid to late eighteenth century, they took it cold and unsweetened.

∻ *The Chocolate Works* ∻

Even though sugar is chocolate's direct opposite in many ways—light in color, sweet, and granulated, rather than dark, bitter, and solid or powdered—the two substances have many things in common. The technologies for growing, cultivating, and extracting consumable cocoa from the bean and sugar from the cane were rudimentary and did not change much until the nineteenth-century era of steam-powered machinery. In addition, both plants flourished only in specific tropical climates, in a band no wider than 20 degrees north and 20 degrees south of the equator, and in areas seldom falling below sixty degrees Fahrenheit.

Cacao trees produce pineapple-sized pods filled with seeds nestled within a white spongy pulp. Once extracted from the pods, the seeds undergo several processes before they even begin to resemble something marketable and edible. They are first fermented at high temperatures anywhere from three days to a week, during which time the seeds germinate, and the process gives them their distinctive chocolate flavor. Then they are dried and roasted like coffee beans.[5] Finally, the beans, now dark brown and desiccated, are winnowed—freed from their husks—making them ready to endure another long set of processes that convert them into, variously, cocoa powder, cake chocolate, and solid edible chocolate.

Like those to prepare sugarcane, the steps to transform cacao beans into a usable product were necessary and, regardless of technological developments, unchanging.[6] Most basic processing occurred on the plantation itself and, as with sugar processing, was labor-intensive. Cacao in "nib" form moved to

European and American cities for further refining, which included separating the fatty cocoa butter from the cacao solids, or "cocoa." Cocoa butter, originally discarded, became an additive that improved the taste and texture of better solid edible chocolates. As with sugar, tropical cacao plantations and European shops existed so far away from one another geographically and intellectually that consumers found it easy to ignore or be naïve about what producers did to deliver the goods, including conquering and colonizing civilizations and instituting the plantation system of slave labor.

For a long time, chocolate belonged to the Spaniards, who introduced it to Europe during their colonial conquests. As a prince, the future Philip II may have been the first on European soil to sample chocolate when in 1544 a group of Dominican friars took Mayan nobles to meet him. Among the gifts they presented were cacao beans.[7] In the late sixteenth century, the Spanish began serious oceanic commerce in cacao beans. This was the last of confectionery's main ingredients to be exploited for international trade. During the same century, Brazil dominated the sugar trade and Italy was perfecting ice storage and application.

Sugar and cacao were not united in a drink until Spaniards balanced the bean's bitterness with sugar's sweetness, just as Europeans did with tea and coffee. Even though sugar and cacao required similar growing conditions, they were cultivated in different locales and remained separate until Spanish colonists, who had easy access to both, combined them. This "process of hybridization," as Sophie and Michael Coe point out, also involved heating the liquid and substituting spices such as cinnamon and black pepper for the "ear flower" and chili pepper Amerindians added.[8] Europeans also devised their own apparatuses to facilitate chocolate's consumption. The chocolate pot resembled a teapot, but with a straight rather than curved handle. Its lid accommodated a *molinillo,* a macelike, wooden device used to agitate the chocolate and water to produce a frothy beverage. Like tea sets, chocolate pots—made of ornately worked precious metals and delicate, gilt-laced porcelain—were status objects, articulating the spoils of colonialism.

In the seventeenth century, chocolate followed sugar's travels through Europe. In the first half of the century, the Spanish courts adapted it to suit their European tastes—something possible now with the increasing availability and popularity of sugar. By the early 1640s, French royalty used it as a palliative, and in 1644, the Roman physician Paolo Zacchia published *Dé Mali Hipochon-*

driaci, which recommended using chocolate as a drug, joining ice and sugar for this purpose. The chocolate drink reached England in the next decade, and in 1655 Oliver Cromwell's forces took Jamaica and her generous cacao, sugar, indigo, and other plantations from Spain, making Britain a powerful international economic force that controlled two of the most desirable commodities in the Western world. Soon after Cromwell's conquest, English merchants began advertising chocolate to the public as a prepared drink or "unmade."[9]

Chocolate quickly became the preoccupation of buyers and sellers alike. For example, Richard Blome, in *A Description of the Island of Jamaica,* of 1678, marveled more at the island's cacao capacity than even its ability to produce sugar. He described "cocao" as "the principle, and most beneficial *Commodity* of the *Isle.*" Although by the time of Blome's writing the British had clear ownership of the island, cacao exports remained still very much "Spanish." Of chocolate in Jamaica, Blome continued, "in time it will become the only noted place for that *Commodity* in the world, which is so much made use of by us, and other *Nations,* but in far greater measure by the *Spaniards* who alone are enough to take of the product of the *Isle.*"[10] The association of the Spanish with chocolate stuck with the substance and influenced its perceptions even among American consumers a century and a half later.

Many people wrote of chocolate's properties as an aphrodisiac. The Spaniards, who mistakenly thought that Montezuma had used it for this purpose, may have launched the myth. With time, the idea acquired more credibility and proponents, adding to chocolate's mystique and lure. "The great Use of *Chocolate* in Venery, and for Supplying the Testicles with a Balsam, or a Sap, is so ingeniously made out by one of our learned countrymen, that I dare not presume to add any Thing after so accomplished a Pen," Dr. Henry Stubbes wrote in the seventeenth century.[11] "It is strengthning [*sic*], restorative, and apt to repair decayed Strength, and make People strong; It helps Digestion, allays the sharp Humours that fall upon the Lungs: It keeps down the Fumes of the Wine, promotes Venery, and resists the Malignity of the Humours," Louis Lémery asserted in his *Traité des aliments* (1702).[12]

While men reveled in the idea that chocolate could aid them sexually, women heard repeated accounts of chocolate's danger. For example, a seventeenth-century Jesuit church in Mexico reported "a number of homicides committed by the Spanish ladies taught by the Indian women, who, by the use of chocolate had correspondence with the devil."[13] Another cautionary tale for women spread in 1671 told of the marquise de Coëtlogan, who "took so much chocolate during her pregnancy last year that she produced a small boy as black as the devil, who died."[14] These associations with chocolate persisted even through the nineteenth century and informed its modern American meanings, especially its dual and highly charged connotations for men and women. For example, an 1859 issue of the *Saturday Evening Post* quoted a comment by the famous French gastronome Brillat-Savarin on Spanish women sacrilegiously taking chocolate in church because they were so fond of the drink they were unable and unwilling to cease the habit.[15]

By the late 1600s, colonists in North America almost certainly consumed chocolate, and it enjoyed more popularity there than in Britain, where the appetite for tea and coffee overshadowed chocolate drinking.[16] The noted Massachusetts judge and merchant Samuel Sewall, for example, mentions chocolate several times in his diary. In 1697, he wrote, "I wait upon the Lieut Governour at Dorchester, and there meet with Mr. Torry, breakfast together on Venison and Chokalatte; I said Massachuset and Mixco met at his Honour's Table." Sewall also frequently gave chocolate as a gift. In 1702, he brought to a "languishing" Samuel Whiting "2 Balls of Chockalett and a pound of figs," probably as palliatives for the ailing man. Three times in 1707, he recorded making gifts of chocolate: a pound to Mr. Little on March 27, another pound to Mr. Gibbs on March 31, and "for Madame Stoddard two half pounds of Chockalat, instead of Commencement Cake," on July 8.[17] Boston apothecaries advertised chocolate as early as 1712,[18] and in 1729 and 1730, ads began appearing regularly in the *Pennsylvania Gazette* offering chocolate and other "East India Goods."[19] By the mid 1700s, Massachusetts merchants were bringing cacao back directly from the tropics as cargo, circumventing its expensive exportation to England and reimportation to the Americas.[20]

The continually increasing consumption rates for chocolate in the Western Hemisphere encouraged concurrent technological developments. Although affordable solid eating chocolate did not make its appearance until the late nineteenth century, Lémery gives an early (and possibly the first) recipe for a solid edible chocolate, which required making a paste of the cacao beans, adding sugar and other flavors to it (including cloves, cinnamon, and "ambergrease"—a substance extracted from the intestines of sperm whales), shaping it "into what Form they please, and then dry it in the sun." He explained: "Chocolate is used in two Ways: It is eaten as it is, or else they make a very pleasant Dish of it, which is much in Vogue, by dissolving it in some Liquor: Common Water is that which is most used, though others will have Cows Milk, into which they put the Yolks of Eggs, that so the Liquor may as it were lather the more, and grow more thick."[21] These instructions did not call for adding sugar and spices. Like loaf sugar, chocolate usually appeared on the market in a solid cake form, bits of which were then grated, ground, or pulverized into a granulated or powdered form to be combined with other ingredients and then further transformed into something consumable. Lémery's recipe describes an "instant" chocolate suitable for eating in solid form with no further manipulation and convertible to a liquid beverage with no additional ingredients.

Technological improvements during the eighteenth century concentrated on improving cocoa's ability to make a good nourishing drink. The English Quaker firm of Fry & Son began producing cake chocolate in 1728, making it easier to dissolve the pulverized stuff into beverages. In 1765, James Baker of Massachusetts and John Hannon, an Irishman, teamed up in America to make ground cacao beans using water power. And in 1776, M. Douret developed a hydraulic grinding machine in France.

Americans became preoccupied with chocolate in the eighteenth and nineteenth centuries. It was "the most restorative of all aliments, insomuch that one ounce of it is said to nourish as much as a pound of beef," the *American Farmer* declared in 1822. "In all countries where chocolate is known, it is esteemed, and found to be a suitable diet for all ages, more particularly for infants, old persons, those of consumptive habits, and such as are recovering from sickness."[22] As with sugar, although desirable, chocolate was sometimes hard to come by, and people sought substitutes. As an alternative to commercial cocoa, *Hazard's Register of Pennsylvania* suggested using grain from *Holcus bicolor,* which would produce "a beverage resembling in color, taste, and many other qualities, the common chocolate." Like the suggestions for growing sugar beets, which began appearing in the press about a decade later, articles such as this tried to provide useful solutions for people who might not otherwise be able to afford the real thing. The author of the *Hazard's* article, for example, wrote: "A single plant will yield seed enough to produce by a second year's crop, a sufficiency to furnish a family of six or eight persons for a whole year, with a good and nourishing beverage, which is supposed to be preferable to tea or coffee."[23] The *Scientific American* published a highly detailed account of chocolate's growth, cultivation, processing, and chemical composition in 1852, explaining that, "Although chocolate is not a daily necessity like tea and coffee, yet the large quantity consumed entitles it to some notice."[24]

Cocoa's growing familiarity to American consumers occurred along with that of sugar and ice, yet it continued to have a signification wholly apart from the other two commodities. Sugar possessed magical properties that could transform almost anything into a pleasure. Ice inverted and subverted nature in a way that titillated people and significantly improved their comfort levels in the summertime. In people's minds, however, chocolate remained entrenched in its colonial and precolonial origins, signifying the tropics, exoticism, and even sex appeal—all characteristics highlighted and mythologized by nineteenth-century American advertisers.

Chocolates in America

The remoteness of chocolate's sources and the labor-intensive processes used at each stage of its production contributed to its expense and its reputation as a highly charged cultural artifact. Fine chocolates defined and articulated points along the continuum between innocence and sexuality, proper and improper sexual drives, reasonable consumption and indulgence. Soft, luscious, and more sophisticated in form, symbolism, and function, chocolates and bonbons did not rely on novel shapes, colors, and flavors for appeal; nor did they cater to children who wanted the immediate gratification of handfuls of sweet treats for their pennies and nickels. Instead, people savored these bonbons one at a time, bit by bit, bite by bite. They were delicate and sensuous, representing romance and sex, and by extension became linked with the appetites of the women who ate them, as objects of their desires. Because of this close connection between soft candies and their consumers, many authors writing in popular literature used attacks against fine candies as their way of criticizing, more generally, the habits of women. Similarly to the process by which commentators attacked the habits of poor children who bought penny candies, advisors warned women against indulging in a pleasurable behavior yet associated them with it at the same time.

For example, an ornate box of fancy chocolates qualified as an acceptable treat when given by a man to his fiancée, yet a plain package of bonbons purchased by a woman who then consumed them herself in the privacy of her own home was not. Although they centered on the same product, the implications of these two acts were very different. The first reinforced the established gender-based power dynamics of relationships that encouraged men to be active providers and women passive receivers. The second, a self-indulgent and transgressive act, called into question the assumptions of the first. To confound things even further, as the nineteenth century progressed and cheaper lines of fine candies reached the mouths of more women, what advertisers saw as marketable and desirable properties for soft candies became the same ones that commentators decried as sinful and indulgent. So bonbons and fine chocolates, at first glance meek and inconspicuous goods, were actually very "promiscuous," as the anthropologist Nicholas Thomas would say, because they adopted radically different meanings and associations depending on the contexts of their production and consumption.[25]

There was a difference between unsweetened powdered chocolate of the Walter Baker variety and the sweetened solid edible stuff that trickled into the American market in the late 1820s. The latter was made possible by technological developments that processed cacao beans into substances more easily assimilated into chocolates and bonbons. In 1828, Coenraad J. Van Houten, a Hollander, patented a process that produced defatted chocolate powder. "Dutching," as it came to be known, not only made chocolate less bitter tasting but also changed its chemical composition, allowing it to assume powdered, liquid, and solid forms. Dutching made possible a solid eating chocolate. The invention of milk chocolate powder by Henri Nestlé in 1867 made chocolates much more palatable to American tastes. Daniel Peter, another Swiss, used this idea to make solid milk chocolate bars in 1879, an innovation successfully taken up by Milton Hershey some fourteen years later. Again in 1879, yet another Swiss, Rudolphe Lindt, came up with "conching," a process that tempered chocolate liquor by rolling and kneading it, making it smoother and more mellow in flavor. "Before Lindt," Sophie and Michael Coe point out, "eating chocolate was usually coarse and gritty; now it had achieved such a degree of suavity."[26]

The French Confection

Before the technological breakthroughs of the later nineteenth century, the only good solid eating chocolate came from French confectioners, who were a highly skilled lot and sold exclusive goods. Bonbons and chocolates qualified as luxuries because of their price, which directly correlated with the ingredients, the time, and the specialization necessary to make them. While the seductive and dangerous qualities of their Spanish extraction still lingered under the surface, chocolates' association with the French became increasingly emphasized. Americans' Francophilia was so avid that it eclipsed even France's own appreciation for its chocolate goods. "Few French people understood chocolate making to have had a distinctive place in French history," Susan Terrio observes. "If France is known historically for a foodstuff, it is bread, not chocolate."[27]

Especially after the American Revolution, European objects and customs acquired a cultural cachet felt to be lacking in Anglo-American goods, and the upper classes, who found French things especially alluring and appealing, came to fetishize them. In July 1826, for example, Philadelphia's *Album and Ladies' Weekly Gazette* described a New Year's Day in France in mouth-watering detail, replete with the gifts men bestowed on women, including bonbons, jewelry, pastries, gowns, artificial flowers, and stockings, noting: "A pretty woman, respectably connected, may reckon her new year's presents at something considerable . . . in Paris it is a custom to display all the gifts in order to excite emulation, and obtain as much as possible."[28] Gratification of wants and satisfaction of needs separated rich from poor, refined from coarse. In his *French Cookery,* a book aimed at the "wealthy epicure," which became popular in America during the 1840s, Charles Francatelli articulated this in writing, saying that his aim was "to treat of cookery as an art by which refined taste is to be *gratified* rather than a coarse appetite *satisfied.*"[29]

Because bonbons were either imported from France or made in American shops by French immigrants, they continued to be special and expensive goods that signified the sensual indulgence considered part of the cultural fabric of France, "as much a status symbol as the French language, the snuff-box, and the fan," according to one scholar.[30] No wonder Henrion could get away with charging a dollar a pound for bonbons in 1828, more than two days' wages for the average laborer.[31] Indeed, Henrion's Philadelphia letterhead even listed his specialties in French, including "Confitures de France et des Colonies," and "Fleurs imité en Sucre."[32] This connection with high society and free living allowed anything labeled a "French" bonbon to command a higher price and more esteem. "To the French belongs the credit of manufacturing the most costly and elaborate confections, but like their cooking these things are not for the every-day life of the every-day people," one writer observed.[33] A well-developed mystique evolved around French culture—and French food specifically—during the nineteenth century. French confectioners benefited from this and garnered reputations as highly skilled artisans whose goods embodied the epitome of sophistication and class, a belief that perhaps lingered because of chocolate's earlier history in France, when the aristocracy limited its use.[34] In his treatise on confectionery in 1864, Henry Weatherley remarked that while the Germans and English were skilled at "rock work" (hard sugar items), "The French confectioners have not developed any striking ideas in this branch up to the present, the national taste keeping them almost exclusively to the manufacture of superior chocolate and sugar bonbons, liqueurs, pastilles, and comfits, in which they excel all other nations, and which are sold at very high prices."[35]

Outlandish prices did not stop the elite from eating bonbons but merely rendered the practice more fashionable; they continued to enjoy them on special occasions—as private gifts or as treats marking the denouement of lavish banquets and dinner parties. The respected Philadelphia confectioner J. W. Parkinson, who ran the exclusive Chestnut Street ice cream saloon, sold bonbons for as much as $.75 apiece in the 1850s. Over a four-day period in the winter of 1853, for example, he charged Mrs. Fisher of Philadelphia $27.64 for bonbons alone.[36] To put this in perspective, common laborers' monthly wages averaged $18.30 in 1850.[37]

By the mid 1860s, the association of fine chocolate with those of French extraction had impressed itself on the popular psyche to such a degree that the "French" label appeared repeatedly, distinguishing fine from cheaper lines of confections. Nearing the first Christmas of the Civil War, advertisers in the *Charleston Mercury* remarked that Kinsman Brothers continued to offer a "varied supply of candies and sweetmeats" "of superior quality," and amazingly, also stocked candy boxes and other novelties "evidently of late French manufacture and importation."[38] Confectioners often offered both "French and Common Mixtures," and ice cream saloons touted "crème glacées."[39]

American manufacturers gradually found themselves defining their goods in terms of French ones. If they could not or did not import French confections, they promised to make superior lines, or to make goods that imitated French confections; but French confections had certainly become the benchmark. In fact, even though technological developments led to increased production of formerly unique confections by the late 1870s and 1880s, manufacturers still relied upon the "French" appellation to distinguish their finer lines of candy, enabling them to charge more. In 1866, Stephen Whitman, who started one of the most successful boxed chocolate and bonbon businesses in America, boasted in a newspaper advertisement that his new chocolate factory used machinery made in France and "not excelled by any other establishment on this side of the Atlantic."[40] Croft and Wilbur, a Philadelphia confectionery manufacturer, listed its most expensive line of candies in 1876 as "French Nougat," which sold for 75 cents a pound. John M. Miller and Son, yet another Philadelphia manufacturer, advertised an entire line of "French Creams and New Fine Confections" in 1876, saying: "Our Imperial Goods, we claim to be of a superior finish, and our Fine Decorated Creams are of the latest French Patterns." N. S. Dickey & Co., a Boston confectionery manufacturer, advertised wholesale prices for its penny goods, which were very cheap. In 1879, French Chocolate Cream Drops—a kind of chocolate bonbon—cost a relatively moderate 15 cents a pound wholesale when buying a 25-pound box, but this was still 10 percent of a blacksmith's daily wages at the time.[41] In contrast, a "real" French confectioner like Felix Potin of Paris, who sold his goods out of New York City, priced his finest French chocolates at $1.17 a pound a decade later, and his specialties, "Boites de coleurs" [*sic*]—chocolates that looked like a set of paints—sold for over a dollar *each,* wholesale.[42] Confectionery, subject to the whims of fashion, donned French labels, so that even mass-producers could reassure their customers that their goods, although not handmade and possibly having no association with France at all, were nonetheless of the finest grade and on the cutting edge.

The Philadelphia Centennial Exposition of 1876, almost bursting with a bountiful selection of the world's latest commodities, played an important role in creating and reinforcing high and low taste cultures after the Civil War. Messrs. Greenfield and Strauss, for example, who had been in business since the 1850s, exhibited a varied line of interesting confections, which they promised were free from adulteration. "The panwork is superior to French *drageés,* and will take the place of the imported goods, so that before long the people will content themselves with only American candies," visitors were assured.[43] Also at the Centennial Exposition, Monsieur Maillard, a famed French confectioner and caterer frequently employed by Mary Todd Lincoln, assembled what the press described as "the most conspicuous article on exhibition." Showing off his craft and creating an ingenious and effective advertisement for his business, Maillard filled a towering display case with his sugary masterpieces, including

> a huge spire-shaped monument of white sugar, nearly fifteen feet in height, and ornamented with historical figures and groups in sugar and chocolate, illustrating incidents in the history of the United States. These include the signing of the Declaration of Independence, the landing of the Pilgrim Fathers, the capture of Fort Ticonderoga, side figures representing Sitting Bull and General Custer on horseback, etc. These figures are all made by hand. There are also two pieces of confectionery: the one a massive vase called the Medicis, of solid chocolate, weighing 200 pounds; the other a book of enormous size, containing 3,000 varieties of bon-bons and chocolates, made by Mr. Maillard and appropriately entitled, "Une Voyage dans L'Isle des Plaisirs."[44]

People literally bought into the idea of exuberant pleasure when consuming things French, which still evoked and signified the decadence of the ancien régime.

There were qualitative differences between handmade French candies and those spat out by machinery, even though the latter might have names like "Royal" chocolates, "Imperial" bonbons, and "French" creams. Yet by the end of the century, when huge confectionery manufactories had overtaken smaller operations, skilled confectioners struggled to stay in business, defending and redefining their niche. They still offered the most luxurious goods, but technological developments increasingly produced imitations of fine French bonbons, which traded on their established cultural currency. Even the conservative *Boston Cooking School Magazine* provided a recipe for crystallized peppermint leaves, "from a fashionable French candy store" at the turn of the century.[45]

⁓ *Boxed In* ⁓

A fancy bonbon box that tangibly embodied the French aura functioned on many levels: it concealed the candies, protected them, and packaged them in something that expressed the taste culture of the giver and designated the purpose of the gift. Chocolates purchased for oneself or to serve in cut crystal candy dishes at a dinner party needed only an inconspicuous, utilitarian plain white box or waxed paper bag. As one candy manual suggested to the retailer, "when people buy for themselves and not for presents, they want candies without the addition of costly packages."[46] However, a present to someone else required an outer packaging of suitable prestige in order to remove its contents, the candy, from the commodity sphere and to suggest the same kind of personalization that formerly accompanied homemade gifts. Bonbons themselves, made of fine ingredients like delicate creams, subtle flavorings, and fragile coatings of hand-dipped sugar, required boxes for protection. When opened, they showcased the candies with complementary tissue papers, filigreed doilies, and gleaming foils. Bonbons' packaging circumscribed their specific contextual meaning and, when of cast papier-mâché, decorated with colorful pictures, and garnished with lace trimming, greatly added to their sentimental value. While manufacturers packed their mass-produced bonbons into plain white or pastel boxes tied with ribbon (sometimes to match the color of the candy inside) that created convenient and uniform packing and shipping containers, fine confectioners nestled their much fancier articles into suitably elaborate packages.

Prepackaged candies relied heavily on the labels of the boxes to identify the product inside and to capitalize on brand reputation and character. Many large candy manufacturers, like the Walter M. Lowney Company of Massachusetts, had their own box factories, too, saving money through vertical integration. Lowney's produced plain boxes but preferred the ones with fancier decorations, "from the best lithographers of Europe and America, employing the best artists engaged in commercial work, and the choice of novel and artistic papers from all countries between Germany and Japan." These elaborate decorations and fancy papers added to the exoticism and uniqueness that one purchased and subsequently bestowed with each gift of chocolates (fig. 20). The boxes also helped create personalities for individual candies and candy samplers: Lowney's, for example, produced the "Crest," the "Cameo," the "Cavalier," the "Betty Alden," the "Debutante," and many others.[47]

Moving directly from factory to confectioner, "jobber," or traveling salesman, bonbons shed their utilitarian shipping boxes when they were placed in trays as part of store window or case displays. This way a confectioner could offer an array of treats both of his own creation and procured from factories.

FIGURE 20. Packaging was often more important than the chocolates inside, and there was a box suited to every occasion. Sparrow's Empress Chocolates. Trade card, ca. 1910. Warshaw Collection of Business Americana. Archives Center, National Museum of American History, Smithsonian Institution.

A customer chose a variety of these candies to place in a box. Selected because of personal preference, the package suggested a uniqueness appropriate for a gift being given to someone else. In an age when few people, especially men, actually made gifts by hand, consumers tried to express the personal by refashioning it in the form of individual choice and taste rather than with handiwork.

The man's choice of costly containers to house the selection of candies, as significant as the selection itself, reflected the depth of his pocketbook and the depth of his sentiment. "The box for the candy is a very important item," noted one trade journal. "It is like the hat to the girl or the frame to the picture. A box given to Prince Henry of Prussia when he was here, a hand-painted satin affair, cost nearly $200 and many a private order costs $70 or $100."[48] England, France, and Germany—the countries that also excelled in handmade confections—made the best boxes, and exported them to American confectioners and box sellers at the turn of the century.

Better confectioners who catered to the elite clientele even opened up "bonbonnieres," separate sections in their shops for the "flattering habit" of men who bought such containers. Often, confectioners accompanied men to their box departments even before they looked at the candies, and boxes imported from Paris carried the most status. "Only the richest silk, satin, brocade embroidery and artistic brush work is employed in the making of the foreign or domestic bonbon box, and for one charming device $35 was asked and received before a single candy had been put in it."[49] Maillard pointed out just how radically a box could affect the price of his confections when he explained in reply to a query "that the quality of Chocolate in the Mignonettes is all the same. Regarding the Prices, they range from 10¢ each to $2.00, in accordance with size and style of package, and with the selection made."[50]

Fancy boxes always denoted special occasions, whether to celebrate a holiday or to mark a special stage of one's relationship with another. For example, P. Arnaud, a French confectioner and box importer living in New York City during the 1870s, charged one of his customers $17.50 for a "Fancy Bonbonniere," while what was inside, two pounds of bonbons and two pounds of fruits, cost an additional $5. Considering that an entire barrel of flour could be had retail for $7.50, this was an important, high-class gift. The entire thing cost $22.50, not including Arnaud's 75-cent charge for express postage.[51] Tiffany and Company even displayed a *bonbonnière* at the 1889 Paris Exposition. Made of sapphire-encrusted platinum and silver, the bauble's inner compartment was large enough to hold but one piece of candy.[52]

"Of course there are plainer boxes to suit more moderate tastes and purses," one trade journal reassured, "and there is thus a pretty wide range of choice so that it makes little difference how critical one's fancy may be or how more modest one's tastes, no one need fail to find a proper selection."[53] By 1900, marketers convinced both givers and receivers that they could effectively express certain emotions through properly selected commodities: "The designs and pictures on some of the best [candy] packages are chaste and elegant, while in the vast assortment of decorative embellishments every taste and fancy may be gratified."[54] The presence of the decorative box transformed the thing itself from an assemblage of commodities selected from an abundance of other similar ones into an individualized, personalized, seemingly unique gift.

Fancy boxes could also become tangible relics—souvenirs—of the moment of exchange, and mimicked the construction of the soft candies within them by being an object with a hard outer layer and a soft center. By extension, boxed candies became metaphors for late-century women, often borrowing the material trappings of their current fashions by arranging frilly layer upon frilly layer around a special core. When presented as romantic tokens, confections served as both symbolic and tangible evidence of the courtship process. Boxes, conflated with emotional sentiment, functioned as vessels of emotion and alluded to people, women themselves. When the candies had been eaten, women used the keepsake box to store hat pins, needlework supplies, or even jewelry.

But more than anything, bonbon boxes embodied personal sentiment that, fused with economic considerations, established a direct relationship between money spent and intensity of emotions felt. A trade journal remarked on the disappearance of the plain white box, noting, "It is a very bold or very careless American swain who would dare send his lady love a gift of candy in this rough guise, and to the traveling foreigner it is a matter of gasping surprise to see the American man cheerfully buying bonbonnieres at the price of a silk hat and send it to a girl on whose heart he has no special designs."[55] To outsiders, American materialism may have seemed odd; yet Americans knew that, in their highly commodified world, sending candies to an intimate in the "rough guise" of the plain white box was a social gaffe of the highest order.

Gifts made of these elaborately presented bonbons and chocolates need to be considered in light of the foregoing. They came to the recipient pregnant with meaning even before the boxes were opened and the candies eaten. "Confectionery gifts can express, strengthen, challenge, or undermine social relations because they represent an important medium through which to communicate social messages," Susan Terrio observes.[56] Many of the nineteenth-century American advice books covering etiquette generally and courting rituals specifically enumerated the rules for giving gifts, and especially circumscribed those demonstrating romantic interest. As highly sentimental objects, presents like finely packaged bonbons were offered only in special circumstances, indicating the degree to which they had accrued symbolic importance.

Popular literature's increasing preoccupations with chocolates and bonbons marks their increasing presence in the culture as a whole, their gradual popularization, and their concomitant use as meaningful gifts. Gifts and commodities may be antithetical,[57] but the age of industrial capitalism endowed them with shared qualities. Fancy chocolates and bonbons became important signifiers of meaning because, beautifully packaged, expensive, and imbued with the French aura, they were simultaneously commodities *and* gifts, vehicles through which people exchanged emotional sentiment. Their emotional symbolism could not be separated from their material worth and the social status they conferred. In fact, these latter qualities made them effective instruments to convey intangible feelings. While etiquette manuals of the time emphasized increasingly controlled contexts for the gift exchange of such rich sweets, advertising kept defining their importance as gifts in relation to their status as commodities. More important than giving the gift was the "proper" selection of its type, appearance, and brand. These gifts of fine candies, then, indicated social prestige, membership in the proper taste culture, and emotional sentiment. In addition, they represented both giver and receiver at the same time. If the nature of the gift generally (jewelry rather than candies, for example) or specifically (penny candy rather than bonbons, plain rather than fancy boxes), was not suitable for both people's situation at once, it signaled in a material, concrete, physically evident way the disjunction of the relationship. The gift could also express this disjunction if it was spurned. People communicated through and established social relationships with the commodities that inundated them.

Fine chocolates and bonbons indicated taste and economic ability as much as they conveyed genuine affection. By the turn of the century, when brand names became attached to middle-brow chocolates, their advertisements stressed to men the importance of purchasing just "the right" box of candies. For example, an advertisement for Schrafft's chocolates showed the silhouette of a couple seated face to face (fig. 21). In it, the woman offers the man a sample from her box of bonbons, and the text advises, "Judge a man by his cigars, a woman by her confectionery. Refined women insist on Schrafft's chocolate bon bons."[58] Another example, from a 1908 issue of the *International Confectioner,* advertising "Sorority Chocolates," read, "The College Girl *Knows* Good Candy," and continued, "The gift of '*the man who knows*' to 'the girl who appreciates.'" Since a woman could discern quality candy from the rest, and might possibly be a better judge of that than of a man's character, he was wise to choose a reputable brand. Accomplishing this, he demonstrated to her not

FIGURE 21. By the turn of the century, mass-produced chocolates continued to position themselves as status goods. Schrafft's advertisement in George Hazlitt, *Historical Sketch of the Confectionery Trade of Chicago* (Chicago: Jobbing Confectioners' Association, 1905), 143. Library of Congress.

only his generosity but also, and more important, his ability to pick the right kind of chocolate—to use his prowess and taste to select the correct candy. The ability to discriminate among seemingly similar commodities was meant to impress.

Gender stereotypes played a significant role in determining how businessmen advertised bonbons and how they wanted the public to receive them. According to popular advisors, it was permissible for a man to give a woman a gift, but only under certain circumstances: "Unmarried ladies should not accept presents from gentlemen to whom they are neither related nor engaged," wrote one author, expressing the common rules of the 1870s.[59] Because gifts reinforced the bonds of blood relations, and initiated romantic ones, their exchange began a cycle of obligations, which had to be nurtured or avoided, depending on individual circumstances. Therefore, advice literature specified proper and improper situations under which certain items could be presented. Mrs. Oliver Bunce, for one, offered the following advice: "A gentleman may send a bouquet, a box of bonbons, or a souvenir of any sort, to a lady with whom he is well acquainted."[60] She was sure to point out that such gifts had to be from a man to a woman, and only between those who were "well acquainted," so that the underlying sentiments would be mutually understood. *Godey's Lady's Book* suggested in 1848 that children exchanging Christmas presents might imitate adult behavior:

> Fruits and cakes, neatly dressed with gold leaves, are divided equally among them; but it is a rare thing for the boys—who always cherish some gallantry for their sisters—not to give up their share to the girls; and new embraces and tokens of tenderness among the little ones, follow the new division. Where the families are large, it often happens that each boy has his pet sister who receives his share of the sweet things, which she requites by some needlework especially adapted to his use.[61]

These boys and girls grew up to be adult (male) givers and (female) recipients of such chocolates and bonbons when they came on the market a decade and a half later. Since gifts, as commodities, involve reciprocity, the power relation constructed here meant an equivalency of material gifts actively given by men—especially sweets—and passively received by women. Boys would get "embraces and tokens of tenderness" in return for their sweets. "With chocolates so heavily identified with the female, giving a man a whole box of chocolates is a heterosexual *non sequitur,* although a woman may play Eve's part, tempting him with one or two from her own box," Diane Barthel observes.[62] A woman was allowed to give a gift to a man that she made herself—a small token of "some needlework" such as an embroidered handkerchief or appliquéd watch fob. But the right to lavish sweets on one's sweetheart was a male one. "Much masculine prestige is contingent on the ability of males to provide luxuries and satisfactions for females," James H. Barnett points out.[63]

Gifts of fine candies also indicated the latest state of the courtship process, defined as "the art of pleasing."[64] Using etiquette books to understand the established forms and standards, one could tell how serious one's relationship was by the kinds of gifts exchanged. Advertisements both created and reinforced these ideas. One was so bold as to assert that similar gift exchanges did not happen much after marriage by asking hypothetically, "When you were engaged your fiancée received a box of Huyler's [candy] almost daily. How often does your wife receive a box of these delicious confections?"[65] Although much of the later advice literature relaxed the limitations on the situations in which confectionery qualified as an appropriate gift, confections packed in fancy boxes were still classified with jewelry, perfume, and other intimate gifts that signified romantic relationships.[66] Fine candies facilitated the courtship process and served as a common language, which was understood as such: "It is not to be denied that a judicious taste in chocolates is a serviceable attribute in a wooer. . . . The frugal swain who should present the penny bar of chocolate broken off ruthlessly from its weather-beaten comrades in the window, deserves, of course, no other fate than which is traditionally assigned to the hindmost." A man courting a woman would give her many gifts as romantic tokens, including chocolates and bonbons, but popular culture advised him to carefully choose goods that conveyed both the proper kind of sentiment and his "judicious taste" and "nice discrimination," because his choice of confectionery concretized his "courteous devotion."[67] The object lesson was that giving generic chocolates was insufficient, for "penny bars" often quite literally cost a dime a dozen, and could only reflect an equivalent valuation of the genteel recipient. At the end of the courting phase, which culminated in marriage (and an equally sugary wedding cake) or disengagement, men no longer employed gifts of confections for romantic ingratiation.

⁖ *Selling Sex* ⁖

Professional confectioners profited from these associations, clarified them, and made them cultural standards through shopwindow displays, gift box iconography, and magazine and newspaper rhetoric. Consumers came to see the fleetingness of fine confectionery as a metaphor for romance. Chocolates, especially, as a "fitting token of the ephemeral passion" of love, were closely linked with courtship and sexuality. "When once the gallant would indite a sonnet to his mistress' eyebrow, he now purchases a box of bon-bons addressed to a destination a few inches lower. Instead of singing [to or of] her cherry lips, he fills them," a confectioner gushed.[68] Confectionery advertising replaced one oral activity with another: sentiments of adoration might be better expressed with sweet treats than sweet words.

Soft-centered candies as private, interior, and intimate forms of confectionery represented richness and sumptuousness, but in a subtle way. Luxury confections were by nature mellow in flavor, delicately colored, and composed of hidden centers wrapped in and protected by chocolate or fondant coverings. Very different edibles from the bright, hard, representational penny candies enjoyed by children, these adult candies targeted refined, sensitive, mature palates—savored by tongues rather than crunched with teeth. They dissolved from the warmth of one's body, like a kiss.[69] The 1905 Schrafft's advertisement of a woman offering chocolates to a man (fig. 21) describes them as "peculiarly delicate," "melting," and "satisfying."[70] The compatible cultural meanings and physical qualities of chocolates and bonbons made advertising's job of reinforcing these connections much easier.

Confectioners realized that they could stimulate trade by touting soft candies as appropriate romantic gifts. Advertising rhetoric played on confectionery's properties both as an aphrodisiac—drawing on its Spanish lineage—and its suitability as a gift: "Confectionery plays an important part in the development of love affairs, and does so in a perfectly natural and innocent manner. . . . Many stories are told of the powerful influence that is exerted on Cupid's behalf by a pound of chocolates."[71] This article continued by telling the story of two American men who successfully sold "Love Candies" through the mail, which they claimed would "enable any young man to win the affections of a lady." A professional confectioner, describing the importance of shopwindows for the female trade, even likened chocolate eating to the original sin

FIGURE 22. The sexually suggestive aspects of chocolate were often overtly expressed. "Strawberry Deserts [*sic*]." Candy box, ca. 1900. Warshaw Collection of Business Americana. Archives Center, National Museum of American History, Smithsonian Institution.

and women's inability to resist temptation: "I sometimes think if the serpent had tempted Eve with a box of modern fine chocolates she would have got much more sympathy for her fall among her sisters of to-day."[72] Because eating something like a chocolate-covered cherry supposedly approximated the sensations of passionate kissing, it constituted a form of oral foreplay. This metaphor appeared not only in trade journal texts but also in packaging and advertising itself. For example, a box that held "Strawberry Deserts [*sic*]," depicted chocolate-covered strawberries in a way that made them look like lips (fig. 22). Shown bright red inside with a rich chocolate coating outside, the candy images themselves were highly sexualized.

These various sexual allusions made chocolates and other soft-centered candies perfect Valentine's Day presents, and fancy boxes of expensive candies became the gift that marked the occasion after the Civil War. People had exchanged paper and lace valentines since the late 1700s; in the second half of the nineteenth century, confectionery accompanied them. Sweets adopted the function of traditional valentines, becoming "the very embodiment of romantic love, tastefulness, and fashion."[73] Three-dimensional equivalents of flat valentines, boxes of candy appropriated the valentine aesthetic and were made of the same materials: heavy cardboard, hand-painted satin, artificial flowers, lace papers, and gold lettering spelling out romantic sentiments.[74] They not only adopted a feminine sensibility but also seemed actually to mimic women: like the female genteel, who lived their lives tucked away under layers of lacy undergarments, corset reinforcements, and silk and brocade dresses, bonbons, too, came wrapped in candy coatings, tissue paper doilies, and lace-trimmed silk or satin boxes.

By the end of the century, enterprising companies mass produced these "hand-crafted" valentines. An 1894 issue of the *Confectioners' Journal* reported candy's gradual replacement of paper valentines as gifts, making clear aesthetic distinctions between the two forms, stating, "Pretty valentine boxes, filled with choice candies, would be a far more refined and welcome gift than the maudlin trash and sentimental nonsense sent as valentines in the present day."[75] Professional confectioners, of course, had a vested interest in promoting fine candies as the newest, latest, most fashionable trend in Valentine's presents and earned a lot from box sales alone. The combined efforts of exclusive confectioners and mass manufacturers ultimately proved effective. To this day, most candy manufacturers remain dependent on the business generated by sales around February 14 for their success.

Acker's completely commodified the holiday by including a complimentary Valentine's Card with their boxes of chocolates: the card, too, was mass-produced and not even selected by the purchaser of the chocolates (fig. 23). What is more, the card, purportedly meant to express a handwritten and heartfelt sentiment, advertised the candies. The two women depicted offer boxes of candies described in the Acker's campaign: "A Heart-Shaped Box in rich colored silk, embellished with ribbon and flowers and filled with our select assortment of bon bons and chocolates, is a luxurious and appropriate Valentine gift."[76] By the time of this campaign, 1903, the sentiments of Valentine's Day had become inextricably embedded in the world of commerce, and a personalized, heartfelt gift merely a purchase away.

What titillated consumers about chocolates and made advisors wary, advertisers interested, and confectioners successful was their association with sexual prowess, which in distilled form created the romantic power of the boxed

FIGURE 23. A mass-produced "personal" gift card. Acker's Valentine gift card. *Confectioner and Baker,* March 1903, 6. Library of Congress.

bonbon. By 1900, there was no doubt that chocolate's reputation as "the apex of the seductive edifice of confectionery" had been cemented, perpetuated through generations of Europeans and Americans since Montezuma's time.[77]

Chocolate companies invested themselves in the supposed origins of their products, and much of their promotional literature detailed the more exotic aspects of its history. For example, in an advertisement for his company, the British chocolate magnate Richard Cadbury, writing under the pseudonym "Historicus," cited Bancroft's *Native Races of the Pacific States* of 1882, which quoted Oviedo y Valdés, a sixteenth-century chronicler of the Spanish colonies, as saying that in Nicaragua, "none but the rich and noble could afford to drink [*chocolatl*] as it was literally drinking money."[78] According to Michael

FIGURE 24. Advertisers continued to play on chocolate's exoticism and supposed sexual power by emphasizing its real and mythic past. "Natives Carrying Cocoa, Fruits, and other Produce, as Tribute; DeBry's 'History of America,'" in Historicus [Richard Cadbury], *Cocoa: All About It* (London: Sampson Low, Marston & Co., 1896), 4a. Library of Congress.

and Sophie Coe, this seeming piece of advertising puffery actually held a kernel of truth: "Each time an Aztec took a drink of chocolate from the brightly painted gourd cups in which it was served, he was, so to speak, drinking real money. . . . Small wonder that its consumption was the prerogative of the elite."[79]

Many ads for chocolates depicted bare-breasted women grinding the fermented nibs or other "natives" harvesting cacao pods, reinforcing textual rhetoric with visual imprints on the imagination (fig. 24). These strategies merely heightened chocolate's allure as an exotic and slightly dangerous substance that suggested the illicit. Seen many times in nineteenth-century advertising, this "imperial primitivism," as Jackson Lears terms it, was as apparent in advertisements for chocolate as it was for patent medicines: "The white man enters the dark interior of a tropical land, extracts mysterious remedies, and puts them to the service of the Anglo-Saxon civilization." Lears goes on to describe the voyeuristic components of such "commercial eroticism," and how products and "native" representations of those products were highly charged and "revealed the doubleness of Victorian respectability: the division between public morality and private yearning may have promoted hypocrisy but also allowed room for fantasy."[80] This division was also at work in the gift of chocolates as well, which publicly operated within the bounds of accepted cultural rituals but privately represented elements of the sexual drive.

While it may have taken a long time for chocolate's eventual arrival on the shores of colonial America, it retained some of its historical roots. Although popular literature glossed over sugar's origins and sources because of the unpleasant realities of the plantation system and the institution of slavery, chocolate's rich, if embellished, history enhanced its desirability. An important aspect of this was chocolate's reputation as an aphrodisiac, even in Montezuma's time, which also helps explain why solid chocolate was so easily linked with notions of romantic love during its period of popularization. The Aztecs "considered the drink to have aphrodisiac properties," Bernard Minifie claims. "Historical illustrations show cups of chocolate being consumed at wedding ceremonies, and in the court of Montezuma, the drink was held in high esteem as a nuptial aid."[81] While the Coes contend that stories of the Aztecs' use of chocolate for erotic purposes actually articulated a "Spanish obsession" (along with finding cures for their chronic constipation), it became part of a common lore that has accompanied chocolate to this day.[82]

The seventeenth-century English physician Henry Stubbes was certain of chocolate's efficacy as a love potion, and his 1682 treatise *The Natural History of Coffee, Thee, Chocolate, and Tobacco* extolled its superiority to the "stimulating agents" used in other cultures:

> If the amorous and martial *Turk* should ever taste it, he would despise his Opium. If the *Grecians* and *Arabians* had ever tried it, they would have thrown away their Wake-robins and Cuckow-pintles; and I do not doubt but you *London* Gentlemen, do value it above all your Cullisses and Jellies; your Anchovies, *Bononia* Sausages, your Cock and Lamb-stones, your Soys, your Ketchups and Caveares, your Cantharides [Spanish fly], and your Whites of Eggs, are not to be compared to our rude

Indian; therefore you must be very courteous and favorable to this little Pamphlet, which tells you most faithful Observations.[83]

In the early 1700s, Dr. Giovanni Bianchi of Rimini included among his many cures for impotence the advice to "take chocolate often with a good dose of vanilla and aromatics."[84] An eighteenth-century European illustration of an amply bosomed woman presenting her husband with a cup of chocolate was accompanied by the following verse:

> I bring to you a special drink from far across the West,
> Although it's nearest loves on whom it's said to work the best.
> Good cheer it always brings, and your full years renews.
> First take a sip, my dear, and I shall presently;
> and know I serve it to you with all the warmth that's due:
> For we must take good care to leave descendants for posterity.[85]

James Wadsworth also versified on chocolate's rejuvenating and sexually stimulating properties:

> Twill make Old Women Young and Fresh;
> Create New Motions of the Flesh.
> And cause them long for you know what,
> If they but taste of chocolate.[86]

Chocolate's reputation for enhancing men's potency continued to be celebrated until Victorian times. Then, when it seemed that women were deriving more pleasure from the substance (aided by sugar), moral arbiters and health advisors spoke out against the stuff because of its overtly sexual nature. Of course, men giving women fancily done-up chocolates and bonbons remained an acceptable practice if following the proper strictures established by etiquette manuals. But because chocolates conveyed romantic sentiments, they could also unleash unhealthy sexual urges if consumed in the wrong contexts, as the conceits of idle women.

Popular writers in the Gilded Age, from suffragists to misogynists, devoted themselves to the "woman question," expressing a general anxiety about the role of women in the private and public spheres. Fiction writers, academics, cultural commentators, and health and etiquette advisors aimed most of their criticisms at the woman who could now enjoy newly democratized chocolates and bonbons. They assumed that the elite woman, already quite used to luxury goods, would demonstrate refinement by controlling her consumption, and hence be more deserving of pleasurable things because she properly appreciated them. But the woman newly exposed to such temptations would be drawn to them not because of a fine sense of cultivation but because of base physical urges—body over mind. Unable to resist, she would quickly fall into a life of dissipation consisting of chronic bonbon eating and novel reading. Eliza Haywood articulated this attitude as early as 1743, warning domestic charges against coveting their mistress's dainties, "by which I mean such things as either are not in the House, or not allowed to come to your Table." Elaborating, she determined class boundaries based on which luxuries were permissible for whom: "It looks silly and childish in a Servant to be laying out her Money in baubling Cakes, Nuts, and Things which she has no real Occasion for, and can do her no good; and no less impudent to presume to touch any thing her Mistress has order'd to be set by."[87] Nineteenth-century printed sources felt no differently. Magazines, from *Godey's* to *Popular Science Monthly* to the *International Journal of Ethics,* told stories of and conducted studies on women's morality as it related to their temptation by "luxurious" things and their spending. Their opinions, which predictably equated high living with the sins of self-indulgence, luxury, laziness, and indolence, found clear expression in women's perceived relationship with bonbons, the sources for and symbols of self-indulgence.

Society's codes allowed a man to give a woman a box of chocolates or bonbons. But if a woman treated herself to these same confections, she was guilty of pleasing herself instead of others. Fraught with all the implications of self-pleasure, this behavior ran counter to her socially prescribed roles as a chaste, nurturing mother and supportive wife. "Sweets require a clear social context and meaning in order to be perfectly proper: they must be shared, exchanged, given, or consumed in a commensual way," Claude Fischler observes. "Thus he or she who eats sweets alone runs the risk of incurring implicit reprobation for indulging in solitary pleasure, one that is not put to any useful, legitimate social use."[88] It was also risky conduct, because according to popular thinking, a woman by nature could not be trusted to control her own desires. The rhetoric surrounding women's consumption of bonbons clearly articulated the concerns that focused on the purity of women's minds, bodies, and souls and the proprietary stance that advisors took in order to help women control themselves. In addition, the rhetoric encapsulated the ambivalence of a culture that at once allowed and even encouraged women to be the main household consumers and its simultaneous anxiety over the power this engendered in them. Of course, confectioners saw these issues quite differently. Realizing that women and their children were their chief consuming audience, they directed their advertising campaigns almost exclusively to these customers' sensibilities.

Popular critiques of women and chocolates paralleled those of children and penny candy throughout the century by focusing first on the physical dangers of eating sweets, then elaborating on spiritual endangerment, and finally combining both, so that the words of popular advisors, couched in the rational terms of domestic science, fused with reformers' ideas that were already preoccupied by larger societal threats. One of the earliest American records of this rhetoric appeared in Margaret Coxe's *Young Lady's Companion,* published in 1839. While designating "sweetness" as an abstract quality denoting a pure inner character and one's degree of moral fortitude, she believed that sweet things, when eaten with regularity, eroded the integrity of one's physical body. Coxe wrote that it was a "positive sin, daily to pamper the body with rich dainties," in large part because it endangered one's physical being, which God had entrusted to one's "stewardship." Like those who called attention to children's many temptations in urban environments, Coxe pointed out the inability of women to moderate their desires when exposed to the offerings of the confectioner, saying: "Many constitutions in our large cities are entirely destroyed, and a great amount of money wasted, by the habit of permitting young girls to stop in their morning walks, to indulge themselves at the shop of the confectioner or cake-maker. Equally, and perhaps more injurious is it, to load the stomach with such unwholesome aliment on each returning night, as do many of the fair victims of fashionable folly." Confectioners relied heavily on women's business for their livelihood, and at the time of Coxe's writing, they catered to the "fashionable folly" of ladies, who were merely "victims," unable to control their appetites for expensive European candies. Coxe herself thought the temptation so threatening to the Christian temple of one's body that she pleaded with women not to "yield to the enervating power of luxurious indulgences." Indeed, weak women had better "meet these and all other insinuating wiles of the great enemy of our souls, with resolution and prayer."[89] It was a sin to indulge in eating candy often, for it wreaked havoc on one's physiology, and by extension jeopardized one's spiritual well-being; only prayer could mitigate the temptations of confections. Even early in the century, when luxury candies were hard to find and accessible only to the elite, advisors disapproved their consumption on moral grounds.

Later in the century, when manufacturers began making chocolates and bonbons on a grand scale, advisors' warnings gained stridency as they clarified the interconnection between food, sexual appetite, self-gratification, and "self-pollution." Masturbation in the nineteenth century threatened the social fabric because people considered it "the ultimate symbol of private freedom and atomistic individualism," Karen Lystra points out.[90] Eating alone and reading alone, then, were both deemed unhealthy solitary pursuits because they quickly led to the sin of onanism. The popular and prolific nineteenth-century health and diet guru J. H. Kellogg shared with Sylvester Graham the belief that "self-abuse" in women was expressed through "dreamy indolence" and gluttony. Outside temptations led women to "perdition": "Candies, spices, cinnamon, cloves, peppermint, and all strong essences powerfully excite the genital organs and lead to the same result [vice]."[91] Insidiously, women usually partook of these external temptations alone, away from reproachful eyes. Such privacy merely encouraged said self-pollution, and threatened to unleash powerful, uncontrollable desires within women themselves. Kellogg continued, "The taste for novel-reading is like that for liquor or opium. It is never satiated. It grows with gratification. A confirmed novel-reader is almost as difficult to reform as a confirmed inebriate or opium-eater. The influence upon the mind is most damaging and pernicious."[92] George Hillard, speaking to the Mercantile Library Association in 1850, reiterated these sentiments, saying: "There is a great deal of trash current in the form of cheap literature, which, like cheap confectionary is at once tempting and pernicious. . . . Especially this is true of the swarms of novels, of English and French manufacture, which come warping upon every eastern wind, most of them worthless, and many of them worse than worthless."[93] Richard Cadbury even facetiously acknowledged reformers' preoccupation with books and bonbons, quoting an early eighteenth-century author who warned his "fair readers to be in a particular manner careful how they meddle with romances, Chocolates, novels, and the like inflamers, which I look upon as very dangerous."[94]

This association between bonbons and novels, interestingly, reappeared in many different popular texts, especially books aimed at boys. In addition to promoting indolence among women, it was alleged, these particular solitary vices threatened to blur commonly understood and important gender distinctions. Women should certainly not glut themselves with bonbons, but it was even worse when boys did it or engaged in similar acts of self-indulgence. Boys who read trashy novels, advisors believed, filled their heads with deleterious material and looked distressingly like the women who filled their bodies with useless sweets. "A large proportion of our novels, especially those published first in our catch-penny papers or magazines and afterwards thrown out, in pamphlet form, in miserable type, upon still worse paper, and sold *cheap,* are of the exciting class," William Alcott wrote in 1850. "Indeed, they are *intended* to be exciting, just as much as pies and cakes and sweetmeats, and confectionery, are intended to be so!" Alcott continued in the same vein, stressing the idea that people developed sensibilities and tastes for these things: "I will even venture the assertion . . . that it is precisely to those who have been trained up to pies, pastry and confectionery, that these paltry productions of a licentious press are particularly adapted, and that it is they who, in greatest numbers, fall victim to their influence."[95]

Sweets had become so prevalent in the culture that they registered not just eating habits, but reading habits as well; people well understood the comparisons men like Alcott were making. William Greenleaf Eliot, another popular advisor, wrote: "One might as well expect to gain strength to his body from sweetmeats and confectionery, as for his mind from works of fiction."[96] Joseph P. Tuttle took the comparison one step further, suggesting that a life of luxury for a young boy, expressed by sweet-eating or book-reading, would irreparably feminize him, alleging that a boy living a life of "luxury" and "indolence," "ends up being effeminate," and "daintily avoids all unnecessary exertion." Putting a finer point on it, he wrote: "You cannot develop a vigorous manhood by stuffing a boy with confectionery and deforming him in a rocking-chair."[97] Women were not supposed to engage in novel reading or bonbon eating because it was indulgent and sinful; boys were not supposed to engage in these activities because it made them like women.

Many nineteenth-century critics of bonbon consumption referred to women's gluttonous appetites and indolent lifestyles and frequently exaggerated the amount of candy they actually ate, measuring the quantities by number of boxes or pounds. These ideas depicted women as out of control, their sexual appetite consonant with and expressed by their eating habits. A graphic

FIGURE 25. Bonbons encouraged licentiousness and dissipation in girls, according to popular authors and the popular press. "Tired Out," *Harper's Young People,* 13 January 1880, 112. Courtesy, The Winterthur Library: Printed Book and Periodical Collection.

example, an engraving from *Harper's Young People* of 1880 entitled "Tired Out," shows a little girl prostrate on the floor, her legs parted, clutching a box labeled CANDIES (fig. 25). Through a mainstream discourse on diet and behavior, American society expressed its collective concerns about the nature of women, their proper roles in the culture, and acceptable forms of pleasure.

Eating bonbons endangered the appetite by filling up the stomach with deleterious sweets and resulted in blemished skin—indulgence made manifest on one's complexion. "The skin is a kind of wrapper or covering, in which the whole body is carefully enveloped," George Horner declared in 1848. "There exists a very close sympathy between the skin and all the internal organs, so much so that one cannot be affected without producing a very sensible impression on the other."[98] Some women ruined themselves by spending "a large portion of their pocket-money" on confections, enough in a year "to furnish a poor family with wholesome bread for a twelvemonth," Margaret Coxe asserted. "They are destroying the tone of their stomach, creating disrelish for nourishing food, and laying too frequently the foundation for premature old age, and a host of maladies."[99]

Since a woman's outward appearance reflected her inward character, and people believed that confections affected one's figure and complexion, they often advised against their consumption. This sentiment continued unabated through the century. "Our dames, although we do not advise them to go to bed nightly on a supper of Stilton cheese and London stout like their English sisters, would, we believe, improve their looks if they lived better," Robert Tomes declared in 1871. "By living better we mean feeding at regular intervals upon well-cooked, nutritious food, instead of wasting their appetites upon cakes, sweets, and other indigestible articles."[100] A woman's appearance concerned reformers, because genteel women functioned as "ornaments for the home." Their own beauty was an outward manifestation of good taste and restraint. In 1884, the author of *Don't,* an advice book for women, warned against the aesthetic consequences of eating too much rich candy: "Don't indulge in confections or other sweets. It must be said that American women devour an immense deal of rubbish. If they would . . . never touch candy, their appetite for wholesome food would be greatly increased, and as a consequence we should see their cheeks blooming like a rose."[101]

A woman who favored rich candies over regular meals transgressed in two ways: she both polluted her body with deleterious substances and tried to conceal her habit. Ranting against a "daily glut of bon-bons and pastry," Robert Tomes declared that the "fastidious dame" who shunned food in public might "on the sly swallow cream tarts by the dozen, and caramels and chocolate-drops by the pound's weight." Such women, "in rejecting publicly beef and pudding, and devouring confectionery privately," were in reality, he charged, "gross, and not dainty feeders."[102] Typical of popular writers on such subjects, Tomes overexaggerated the amount of candy a woman could reasonably consume "on the sly"—*dozens* of cream tarts and *pounds* of chocolates. He also castigated the woman who consumed in private, by herself and out of public view: there was something unseemly and suspicious about someone who would indulge herself alone and secretly, free of the controlled and controlling gazes of others. Glutting herself in private on sweets, she transgressed because she took it upon herself to satisfy her desires, and by doing so in private she subverted the culturally established behavioral codes meant to control those behaviors. Through the discourse on confectionery that circumscribed women's behavior, popular advisors could also indirectly discuss proper sexual mores and conduct in general.

"To merely pander to the appetite is sensual and debasing . . . *food is a moral agent,* and has its effect upon every character," the *Boston Cooking School Magazine* argued in 1896.[103] Seen from this perspective, the ingestion of this "moral agent" therefore had to be controlled, and foods like bonbons that gave great pleasure but lacked nutritional value worked as agents of moral turpitude. Because bonbons, containing a lot of sugar, were stimulating but also potentially addictive, and because women in particular had a fondness for them, it made sense to locate the center of discussion about morality at the moment of the bonbon's consumption. John Lewis, a successful New York grocer, could have written about any number of goods and displays he saw at the Centennial Exposition in 1876. Inspired by the many confectionery exhibits, however, he reserved commentary for what he saw as a uniquely American phenomenon, saying: "The candy stores are among the gayest and most prosperous, & the consumption of their wares by *women* & children all the year round is enormous & is often spoken of by writers as a great national failing. If a woman goes out she must not forget a pound or at least ½ a pound of candy for baby—and herself."[104] Women both gratified their own appetites by buying sugared goods by the pound and influenced the diets of their children, in some ways threatening the moral fabric of the next generation.

Bonbons and other fine confections came to be the focus of many entreaties because they were luxurious, ephemeral, and fancy items that like gratuitous sex, masturbation, and novel reading had no purpose other than to take up time and to gratify indulgences. "All personal luxury springs from purely sensuous pleasure. Anything that charms the eye, the ear, the nose, the palate, or the touch, tends to find an ever more perfect expression in objects of daily use," Werner Sombart observes. "It is our sexual life that lies at the root of the desire to refine and multiply the means of stimulating our senses, for sensuous pleasure and erotic pleasure are essentially the same."[105] It is no wonder, then, that social commentators fretted about women's relationship with candies. By the end of the nineteenth century, women undeniably constituted the key market for soft candies and often had the ability—through a household allowance—to buy them for themselves. But popular criticisms articulated cultural contradictions: although confectioners encouraged them to buy sweets, and people eventually came to laud women as "sweet," it was simultaneously feared that once unshackled, women's gustatory appetites, like their sexual appetites, might spin out of control and be the ruin of their households and, eventually, of the entire country.

An interesting counterpoint to this trajectory of chocolate was how it became transformed into a product fit for children. Although tales of chocolate's aphrodisiac properties clung to it throughout the centuries, transferred from the beverage Montezuma was said to have drunk to the solid substance Victorians found problematic, chocolate drinks were stripped of these associations and became linked with nutrition and children. For example, a 1904 advertisement for Baker's Cacao des Azteques chocolate, a product whose name combined Mesoamerican exoticism with French flair, described it as a compound "composed of the best nutritive and restoring substances, suitable for the most delicate system. It is now a *favorite breakfast beverage for ladies and young persons,* to whom it gives freshness and *embonpoint.*"[106]

Writing on chocolate in the 1890s, Alfred Crespi had it on good authority that chocolate beverages were beneficial: "Dr. Carter Wigg suggests that among the articles which might often with conspicuous advantage be given to young children, cocoa should take a high place, as it is exceptionally rich in nutritious properties." However, he went on to suggest, "cocoa prepared for children should be weak, with abundance of milk, or it should be added to other food."[107] As chocolate in its various forms came to occupy a greater presence in the diet, its uses and cultural meanings also became more complex. Until Hershey completely democratized solid chocolate and made it decent enough for universal consumption in the late 1890s, only powdered chocolate diluted with the addition of sugar and milk qualified as an appropriate drink for children, while richer, solid chocolate was seen as suitable for the palates and digestive systems of adults.

Obviously, the companies producing baking chocolate and cocoa—two forms of unsweetened chocolate that had to be incorporated with other ingredients to make them edible—touted the properties of their products as more utilitarian than esoteric. A late nineteenth-century pamphlet published by Walter Baker & Co., for example, asserted that "as a beverage for the table, a mistake has been frequently made in considering chocolate merely as a flavor, an adjunct to the rest of the meal, instead of giving it its due prominence as a real food, containing all of the necessary nutritive principles. A cup of chocolate made with sugar and milk is in itself a fair breakfast."[108] Consumed in this way, cocoa was not a luxury but a staple in poorer households. Combined with flour, eggs, and sugar, it created the foundation for baked goods, also more substantial foodstuffs. But these forms of chocolate and cocoa remained materially, aesthetically, and symbolically removed from pure, solid "chocolate chocolate."

Children—more than men, who were used to their recreational tobacco—had since the 1840s satiated their collective sweet tooth with an array of inexpensive penny candies. Mass production at the end of the century of chocolates like Hershey's, Whitman's, Schrafft's, and Huyler's required more chocolate consumers. To suit children's tastes, manufacturers made chocolates lighter-colored and more delicately flavored. Focusing on children's preference for lighter, sweeter chocolate, Milton Hershey and others perfected the manufacture and selling of milk chocolate. (Hershey's dark bittersweet penny bars were not as popular.) Hershey began his chocolate-making enterprise in 1893 after purchasing the requisite equipment from German chocolate makers who exhibited at the World's Columbian Exposition in Chicago. He revolutionized the market for solid eating chocolate by producing inexpensive goods that did not resemble luxury chocolates at all.[109]

Hershey claimed that the idea of using milk in his candy came from working with familiar caramel ingredients rather than being influenced by the Swiss, who were clearly experts at milk chocolate production. While this may have been so, he used their conching and milk chocolate techniques to perfect his trade—techniques responsible for his long-standing success in the American market. Locating his factory in rural Lancaster County, Pennsylvania, he enjoyed ample and ready supplies of fresh milk and cream, which contributed to his chocolate formula's success. And because Hershey had been making penny candies for decades when he began his chocolate operations (he had once run a candy shop in Philadelphia), he knew how to satisfy children. In them, he saw a ready and eager market for solid chocolates, if he could tailor his goods to younger sensibilities. When Hershey reevaluated his products at the turn of the century, he expanded his milk chocolate line to better serve children's tastes. His candies incorporated the logic that milk added to chocolate made it "good" for children, in contrast to dark bittersweet chocolates, which still counted as "real" chocolate, meant for adults. In fact, one writer in 1905, who decried the popularity of the milk chocolates popular in the United States, compared them to "*chocolate* chocolate," which was not adulterated with something that made it taste "as though it had been made in a cheese factory."[110]

Not only did Hershey alter the flavor of his chocolates by adding milk, but he also changed the shapes of chocolates to appeal to the visual preferences of his little consumers and to make them look more like penny candies. By 1899, Hershey was producing over 130 different chocolate "novelties" in addition to plain penny bars and cocoa powder. They included an entire line of chocolate cigarettes: "Le Chat Noir," "Tennis," "Opera," "Smart Set," "Recreation"; chocolate cigars: "Banquet," "Hero of Manila," "Pride of Lancaster," "Sweet Monopole," "Cuban," "Exquisitos," "Perfectoes," "Little Pucks"; and assorted novelties: "Columbian Gems," "Lady Fingers," "Sweet Peas," "Glories," "Bijous," "Vigoes," "Little Jokers," "Lobsters," "Midgets," "Sweet Peas," "Ozark Sticks," "Zooka Sticks," "Mohawk Sticks," and "Hershey's Chocolate Manhattans." Employing all these novel shapes and imaginative names made Hershey's solid milk chocolates the favorites of children whose parents had grown up playing with penny candy "toys." Hershey also endeared himself to the American public at large by creating domestic chocolates that were within the reach of everyone. In fact, the Hershey's nickel bar, which fluctuated in size, remained the same price from 1903 until 1970.[111]

Other manufacturers quickly appropriated Hershey's marketing strategy and began producing their own lines of small, affordable solid eating chocolates for children. Hawley and Hoops, for example, also created a line of chocolate cigarettes, pipes, and cigars, aiming their advertising directly at those too young to afford or be allowed the real things (fig. 26). By 1901 the *Confectioner and Baker* celebrated the profusion of common chocolates, all—like the hard penny candies—replicas of things found in the natural world and in the marketplace. As favors, these chocolates also joined another growing institution—the middle-class child's birthday party, where they contributed to the theme, be it gardening (with miniature hoes, rakes, and spades), transportation (small chocolate locomotives), or animals (the chocolate zoo). Hershey's also followed accepted gender constructs by offering goods specifically geared toward boys or girls, as was the practice with penny candies. To girls, the company offered dollhouses with furniture and miniature sets of dishes made of candy. Boys could buy "military or naval outfits" or replicas of fire engines "designed to enchant the coming man." With the help of Hershey's and other companies, by the turn of the century, the chocolate vehicles that conveyed

FIGURE 26. Chocolate allowed little girls to transgress age and gender barriers. Hawley and Hoops, New York. Trade card, ca. 1890. Warshaw Collection of Business Americana. Archives Center, National Museum of American History, Smithsonian Institution.

romantic intentions had been desexualized—reshaped into whimsical forms and drained of potency—in order to appeal to and be more suitable for a younger and more innocent audience. The only excuse for adults to eat such playful chocolates, according to Hershey, was if they temporarily "turned child again."[112]

Yet even Milton Hershey could not fully escape nor resist the lingering associations of chocolate and romance. While running his Philadelphia confectionery store in the 1870s and 1880s, he had developed one of his specialties, "French Secrets." Each French Secret consisted of a slip of paper inscribed with a romantic verse wrapped around a piece of hard candy, all hidden under a layer of tissue paper.[113] At the inception of his chocolate works, Hershey revived his old idea, manufacturing "Sweethearts," drop-shaped chocolates with hearts embossed in their bottoms. The Hershey's "Kiss," born in 1907, combined the concepts behind Secrets and Sweethearts, resulting in a product that is still wildly popular today. The romance of chocolate made more innocent for the consumption of children materialized in the Kiss. Hershey's advertising also emphasized this aspect, frequently showing a boy and a girl about to share an innocent peck.

Manufacturing confectioners also capitalized on children and chocolate during the holidays. Valentine's Day was reserved for the sweet language of lovers, but manufacturers created chocolate Easter rabbits specifically for children. Leigh Eric Schmidt has described the commercial cooptation of formerly religious holidays during the nineteenth century. Easter was no exception. By the late 1870s, "the number of goods for Easter—cards, toys, plants, flowers, confectionery, and other novelties—mushroomed at an astonishing rate." Merchants and mass production took over "the traditional links that had joined Easter, nature, and folk culture."[114] This was part of the trend that replaced human relationships and personal sentiment with commodities: "One of the features which marks the American celebration of Easter as distinctive from that of other nations is the universal candy eating," boasted one trade journal.[115] In 1902, the *Confectioner and Baker* even claimed that 50,000 candy Easter eggs were being produced for the Chicago market alone, and reassured its readers, "So long as the real egg continues as an Easter symbol the candy egg promises to be a confectioner's staple for every season."[116] A *Scientific American* article on Easter candy written in 1906 marveled at the "remarkable development of modern manufacturing activity" derived "from popular customs, often national, though more frequently religious" which were originally "largely individual or personal in character."[117] Chocolate Easter bunnies, along with marshmallow chicks and sugar eggs, exemplified the commodification of religious holidays. Even the people living in rural Lancaster County, Pennsylvania, ate chocolate rabbits and eggs at Easter. Cream eggs, soft like Valentine's chocolates, encouraged secret consumption and slow savoring. These chocolate eggs, "were not the elaborately decorated ones advertised by the confectioners, but small chocolate-cream ones with, perhaps, a yellow center to represent a yolk. These were usually hidden away for slow consumption, a thin slice being whittled off each day, to make the pleasure last as long as possible."[118] How children ate chocolate cream eggs at Easter resembled the way women ate soft-centered candies on Valentine's Day. "There appears a tendency for Easter culture elements to develop and proliferate around children and women, especially in the elaboration of the secular phases of the pattern," one sociologist notes.[119]

Merchants found the Easter rabbit to be a perfect holiday icon, because it was so easily recognizable and its anthropomorphic nature readily appealed to children. Further and further divorced from its ties to agrarian culture and the changes of the seasons, the Easter bunny turned out to be yet another character implicated in the story of commerce. "Soon, for instance, it was not enough to have an ordinary rabbit; instead, the confectioner's hare itself became a creature of fashion and oddity. It played a drum or was put in harness like a horse or was dressed up in the latest styles. . . . they themselves, in other words, had become model consumers," Schmidt notes.[120] Chocolate and stuffed rabbits served as children's toys and treats, teaching them in turn how to be "model consumers," and so converted what had originally been a deeply religious holiday into one whose terms were defined by individual and mass consumption. Chocolates in fanciful shapes and diluted forms suited the tastes of children well, because they were far removed from sexually charged boxed chocolates.

Industrialization

Once geared up, manufacturers ably supplied the market with a steady stream of chocolates. Yet they feared that business from women and children would not be sustainable unless they completely removed the taint of sinfulness from their products. Hershey and others had made their chocolates literally and figuratively more palatable to the younger set, breaking down psychological barriers to consumption. Appropriating the rhetoric used a decade later by pure food advocates, companies demonstrated the purity of their goods by stressing the virtue of their workers and the cleanliness of their factories. Many large chocolate operations, including the British companies Cadbury and Rowntree, were run by Quakers. Milton Hershey applied his Mennonite tenets to his factory system and created a wholesome company town in the process. One writer praised the environment in Cadbury's female-populated packing room, saying, "The most perfect cleanliness prevails. The half-score or more of girls, who work under the superintendence of a forewoman, are all dressed in clean Holland pinafores—an industrial uniform."[121] This same writer easily shifted his analysis from issues of physical cleanliness to morality, stating that "commercial progress cannot well be considered apart from moral progress; we want to know not only how work is done, but who and what they are who do it. . . . We may therefore very properly say a few words respecting the *employés* in the cocoa-factory. No girl is employed who is not of known good moral character."[122] Factory owners reassured people that the goods they purchased came from spiritually enlightened hands and were therefore pure at point of origin; it was only by way of the context of consumption that chocolates could become otherwise.

Soft candy manufacturers who were by the 1870s slowly encroaching upon the business of the handworkers, endeavored to make the best candy possible in large quantities, which they had a ready market for in the middle classes. Even in highly mechanized manufactories, fine candy production still required skilled and semi-skilled handwork at many stages, from demolding and dipping to wrapping and packing. Furthermore, since external factors like the weather affected the behavior of the raw materials, production demanded experienced workers to oversee each batch and ensure product consistency. In addition, molding bonbons into various forms necessitated workers who prepared the trays of cornstarch by embossing candy shapes into them, gently pouring in the molten candy, and being able to demold them when cool. Even with the introduction of machinery for these processes, which did not occur until the 1890s, their production still required the skillful eyes and hands of an actual person. Finally, and most important for profit margins, finishing the bonbon centers by dipping them in different flavored and colored coatings necessitated careful and consistent handwork because outer appearance ultimately identified the bonbon as a fine candy, and signified that, unlike penny candies, they were not generated by fully mechanized processes. Even after full mechanization, bonbons and penny candies remained in different "classes" because bonbons retained the illusion of being handmade and thus somehow transcended their existence as one of a number of similar commodity forms.

Candy factory managers divided the labor by gender, so that men handled the larger, heavier, and usually more dangerous tasks, such as stirring vats of boiling sugar, hauling supplies in and out, and delivering candies in different stages of manufacture to their next destinations. In contrast, women performed the more tedious but no less skilled tasks of finishing the candies by dipping their soft centers in chocolate coating or by packing them in boxes: "The production and boxing of chocolate-covered goods in all their variety of substance, form and decoration entail much handwork, and are the greatest labour-absorbing items in the cocoa and chocolate industry. They explain also the high percentage of feminine labour employed therein," stated one article.[123] The intensity and amount of hand labor justified soft-centered candies' expense, and the trade journals explained, a bit defensively, "why candies made of sugar that costs five cents a pound and cream that costs twenty cents a quart should retail at fifty or sixty cents a pound."[124]

Other people profited in less direct ways by publishing books about how to make confectionery in one's own home. This represented the final push toward democratizing bonbons. Instructional books removed fine candies from the realm of the exclusive confectioner who loaded his store with imported and handmade goods and brought them into the home. From the 1890s to World War I, publishers generated a slew of accessible cookbooks explaining how people could make their own candy. The existence of these books acknowledged yet tried to minimize class differences by making such luxuries more widely available, targeting lower-class aspirations for fine goods. Poorer people still desired this candy, but could enjoy it only by producing it themselves. James S. Wilson aimed his 1904 book *Modern Candy Making* at those "who, from stern necessity, cannot afford to indulge in the best chocolate creams at a cost of from 60 cents to a dollar a pound." Rather than have a class of Americans be "denied," Wilson wanted to put "within their reach the choicest of confections" but at less cost by providing home recipes to make "the most dainty of luxuries."[125] But the only things being made accessible in these books were the recipes themselves. One still needed the time, skill, and supplies to make fancy candies, and while this may have brought more sweets into poorer homes, having to make your own luxuries made them, in fact, no luxuries at all. It was a contradiction in terms.

Mass production combined with the decreasing cost of sugar and the creativity of enterprising minds helped increase the country's taste for and discrimination among sweets. In 1905, the confectioner George Hazlitt wrote: "Where thirty years ago, nearly half of the stock of the jobber [sales agent] consisted of stick candy, it now consists of such goods as dipped walnuts, dipped caramels, nougats, Italian cream bonbons, cream patties, chocolate drops, etc., in fact very nearly if not quite half of the stock generally carried by jobbers is fine chocolate goods."[126] Mechanization both in cacao processing and in candy manufacturing made former luxury goods widely accessible by the end of the century.

Because one of the things that made soft-centered candies so expensive to make and to buy was the outer coating, which required hand-dipping, manufacturers eagerly sought out mechanical solutions to this problem. They searched for a machine that would both put an end to the costly handwork and also continue to imitate its physical characteristics, mass-producing goods that still resembled the finer ones and commanded higher prices in the stores. By the turn of the century, candy companies introduced a machine, suggestively named the Enrober, to coat the sugar fondant centers, thereby replacing a room of fifteen dipping girls with only three to attend to the mechanized operation.

Even though sanitary machines rather than morally pure girls were now producing chocolates and bonbons, the candies remained connected to and surrogates for the human body. One author described the transformative properties of the Enrober, saying, "Naked and unadorned centres are fed on trays to these wonderful machines, and in passing over a vessel filled with liquid chocolate they are clothed in a chocolate coat."[127] Although now a product of machinery, the bonbon was still defined in terms of the human form, its essence being merely a "naked and unadorned" body that was then "clothed" via machine by chocolate. The candy's outer coating, like clothing itself, at once protected and concealed, was ornamental and symbolic.

Half a century earlier, the Philadelphia confectioner J. M. Sanderson had expressed an awareness of technological things to come. In his 1844 treatise on confectionery, he provided detailed instructions about adding decorative piping to bonbons by hand. He valued the artistry of handwork, which could be matched only with the use of molds, an early technique that generated multiples: "Some of the bon-bons, which may be seen in the shops, are proof of what I assert [about beautiful handwork]; and many things are so cleverly done, that many persons would believe that they were either formed in a mould or modelled."[128] But some fifty years later, in an age that celebrated yet feared industrialization, the Enrober's ultimate purpose to add decoration to fine bonbons, like mechanization itself, remained ambivalent and, at times, unclear. For example, the National Equipment Company, a manufacturer of candy-making machinery, claimed that its Enrobers could imitate any design accomplished by hand, "limited only to the ingenuity of the operator." The NEC's catalogue expounded upon the advantages of its Enrobing system, saying: "Chocolates so decorated can be told at a glance. Their decorations are distinctive and they possess individuality. They are fast becoming universally recognized as the decorating mark of chocolates that have been produced, coated and decorated under the most sanitary conditions."[129]

A machine that could coat thousands of candy centers an hour still needed to produce goods with "individuality" to satisfy the wants and habits of the marketplace. Thus machines like the Enrober produced multiple, faceless commodities that, ironically, were regarded as appropriate gifts expressing intimacy, emotion, and personalization.

If this was not contradictory enough, the NEC's catalogue went on to describe other advantages to its Enrobing system, which not only coated the candy centers mechanically, but endowed their surfaces with a variety of designs. Wires of different shapes gently touched the tops of freshly coated centers; when lifted, signature impressions remained in the semi-soft, warm candies. The NEC claimed that its Enrobers excelled at producing candies that reassured the public of their birth from machine, rather than from the unsanitary hands of dipping girls—a valued feature of commodities, especially edible ones. However, one of the main attractions of semi-soft candies was their inherent symbolism, which stood for and yet replaced the sentiments and sophistication of professionally handmade goods. "Many of our most expensive and attractive chocolates today are Enrober coated and wire decorated. It is impossible for the consuming public to distinguish chocolates so marked from those formerly dipped and strung in the old unsanitary way," the NEC catalogue said.[130] People wanted to possess and exchange handmade goods. But they also wanted these goods to be affordable, sanitary, and accessible.

Most significantly, manufacturers and popular critics alike continued to associate chocolates and bonbons with feminine qualities. Physically, good candies and good women were soft, dainty, delicate, and pleasing to the senses. From an economic standpoint, women continued to be reliable consumers of these candies, and therefore advertisers continued to direct their sales pitches to them. At the same time, commentators assumed that women remained irresponsible consumers and urged men to do the buying for them, to control their pleasure. Chocolates in their late nineteenth-century incarnations may not have been considered potent aphrodisiacs as they once had been, but society's strictures regarding the consumption contexts for these confections allowed men to remain the potent buyers and givers and expected women to be the passive, dependent recipients. What is more, to be suitable for children, chocolate had to be—and was—diluted with milk, whether eaten in solid or liquid form. Issues of gender and power continued to express themselves through the language of confectionery, and they became even more overt when it came to saccharine wedding cakes.

CHAPTER FIVE

The Icing on the Cake

∴ *Ornamental Sugar Work* ∴

In 1765, John William Millers came to Philadelphia looking for a job. Claiming to have worked for royalty, and educated in the flamboyant art of sugar sculpture, he most effectively publicized his skills through demonstration rather than rhetoric—showing, rather than telling, potential patrons what he could do. In June, Millers placed a large announcement in the *Pennsylvania Gazette* bidding the public to see his masterwork, at the cost of one shilling each:

> This Piece of Art represents a gigantic Temple; on the Top of which stands Fame, with both their Britannic Majesties Names in laurel Wreath; in the Temple is the King of Prussia, and the Goddess Pallas; at the Entries are placed Prussian guards as Centinels; without are Trumpeters and Kettle drummers on Horseback, inviting as it were, the four Quarters of the Globe, who make their Appearance in triumphal Cars, drawn by Lions, Elephants, Camels and Horses; together with many more magnificent Representations, not enumerated in this Advertisement.[1]

If people came away feeling that "the Work is not of such Art as mentioned," Millers promised to return their money. A monumental piece of artwork, the sugar temple displayed Millers's skill and imagination and brought glimpses of royal lifestyles to the American populace. He also offered to make, "at the most reasonable rates," "finer Pieces for Gentlemens Tables, Weddings, or other Entertainments."[2]

Rather than using refined sugar to augment recipes, to add flavoring, or to supply a needed energy source, professional sugar workers used it as their main medium to create pieces of art. While not confections to be eaten, sugar sculptures were confections for the eye, to be consumed visually and "tasted" vicariously. Sugar in its most luxurious and least utilitarian form, the sugar sculpture clearly articulated bourgeois values, which eventually found wider popularity in wedding and other cakes, ornamental objects in confectioners' display cases, and even amateur homemade efforts. These pieces demonstrated the skill, ingenuity, and art of the sugar worker. As status symbols, they stood in confectioners' windows, on elite banquet tables, and at important ceremonies for all to see. Because pieces of ornamental sugar work remained appreciated yet uneaten, they negated sugar's possible role as a nutritive source and merely underscored its symbolic dimensions. And although by the end of the nineteenth century, their form had become highly feminized, they remained the kind of confectionery that most readily referred to masculine power.[3] Wedding cakes remained among the only enduring saccharine monuments to male power. Eventually, like other handcrafts at the turn of the century, even fine sugar work incorporated mass-produced elements, blurring the distinction between upper- and lower-class commodities and making wedding cakes and similar ceremonial confections into familiar symbols.

⁖ *Food as Art* ⁖

Confectioners who made sugar sculptures in America borrowed directly from the works commissioned by European royalty. Ornamental sugar work continued an age-old tradition of turning food into art, which in modern history became popular at the banquets of medieval French and English courts in the fourteenth and fifteenth centuries. "Subtleties" (also "sotelties"), or *entremets* in French, were dishes served between main courses to both provide entertaining gustatory respite for diners and also to give servants time to clear the table of the remnants of some eight or ten dishes. These edible spectacles, both sweet and savory, became increasingly more elaborate as hosts and their confectioners engaged in culinary one-upmanship: "The best medieval feasts united the sumptuous with shocking, 'unnatural,' and incredible events. . . . Cooked peacocks were served with their iridescent feathers. Tethered live birds were baked into pies; the crust cut, they sang."[4] Sometimes fanciful and often outrageous, subtleties served as grown-up toys and playful diversions from otherwise serious eating.

Through time, subtleties transformed from relatively unimportant dishes into elaborate confections—visual amusements presenting allegorical motifs that "contributed to the ceremony of the hall and its theater of feasts."[5] Although made out of ingredients such as spun sugar and marzipan paste, they were more often admired as impressive artistic accomplishments than eaten. Miniature yet monumental forms, subtleties gave the banquets an air of conviviality and cultural prestige based on power and wealth; in addition, they often communicated political and religious symbolism, often lasting much longer than the banquet itself. For example, at a sixteenth-century banquet hosted by the English cardinal and statesman Thomas Wolsey, he presented an entire chessboard made of "spiced plate" (sweetmeats) to one of his visitors, who requested that a case be specially made so that it could be transported back to France unharmed, as a souvenir or trophy, perhaps.[6] Subtleties tangibly commemorated the occasion of a feast or banquet and lasted longer than a comfit. At the moment presented, they reconfirmed the reputation of the chef who oversaw these banquets and became, in addition to the purveyor of elaborate feasts, an artist who performed his skills for the enjoyment of the banqueteers. Reciprocally, the host improved his status as patron by delivering the chef's abilities to the feasting table. Pieces of the displays taken back to one's home functioned as powerful keepsakes of the event and circulated avant-garde culinary fashions among the very wealthy.

⁓ *Early American Sugar Work* ⁓

When Europeans migrated to America during the seventeenth and eighteenth centuries, they brought with them their customs, including culinary practices. The colonists needed not only blacksmiths and farmers but apparently a few fine confectioners as well. Early sugar workers set themselves apart from other craftsmen and cooks by touting their general skills at pastry making and cake baking and emphasizing their talents at forming sugar into sculptures, the height of an artisan's abilities even in provincial America. One such confectioner, Grace Price, who had catered to and "lived in some of the best Families" in Philadelphia, advertised, for example, in the *Pennsylvania Gazette* (8 March 1758). Because people linked artistic skill with social status, there was call for a person like Price, who could cater a private party or a large banquet in the "genteelist and politest manner," her ad promised. She also touted her abilities to work in both English and French "tastes," indicating the degree to which she and her patrons had absorbed and traded on contemporary signifiers of refinement.

A sugar worker aspired to awaken visualness and whet one's appetite by making products with eye appeal from a coveted foodstuff. The significance of sugar sculptures lay in their novelty and decadence. They negated sugar's practical uses as a flavoring agent and energy source (for the pieces were rarely eaten) and emphasized a confectioner's artistic skill at using such a precious commodity as a sculpting medium. Like the other forms of early confectionery—comfits, preserves, glacés—examples of ornamental sugar work found their way only to elite tables and marked important social events. George Washington used his own French confectioner to furnish an inauguration celebration in 1789, a man whom the president's personal secretary described as "a complete confectioner [who] professes to understand everything relative to those [table] Ornaments."[7] Washington's use of a fine confectioner in this capacity indicates that America's aristocracy had not rid itself of European affectations. His inaugural decorations did not differ in form and symbolic import from subtleties.

Inspired by creations they had heard about or seen, eighteenth-century women imitated them on a smaller scale, using sugar molds and supplementing confected figures with nonedible figurines. "With such porcelain trinkets hostesses could create a table scene resembling Governor Tryon's when on gala occasions his table was set with Italian temples, vases, china images, baskets, and flowers. They could also imitate Martha Washington's table at the official mansion in New York, where in 1791 Senator William Maclay of Pennsylvania saw it 'garnished with small images and . . . flowers.'"[8] Table decorations of sugar, glass, and porcelain provided fanciful figural counterpoints to the food being served. Culinary accessories, they complemented the menu and recalled medieval French *entremets.* Also called "toys," these figurines functioned as adult and highly cultured versions of children's penny toys and candies—

more mature and sophisticated forms that publicly proclaimed status by presenting a profusion of precious whimsical objects meant only to entertain and impress.

Erecting Sugar

Ultimately, the elite's aesthetic sensibilities informed the successful sugar worker's repertoire, because his sculptures had to reflect them: examples of Veblen's conspicuous consumption if eaten; conspicuous waste if not. Also "caterers," confectioners provided their customers with what they wanted—tasteful, amusing, enchanting decorations that pleased the eye and excited the palate. Certainly, making ornamental sugar work was not for amateurs. An issue of the *American Messenger* from 1837, for example, recommended that women should only attempt simple jellies or custards at home. For anything fancier, the writer advised "having recourse to the best confectioner for ornamental dishes." The author continued, "They will cost no more than if made at home, and one or two pretty dishes, properly prepared, will do more credit to the mistress of the house, than half a dozen failures."[9] Selecting the right confectioner and the proper ornamental dish reflected a woman's cultural discernment, an important part of a domestic role that evinced high European affectations. Women's nineteenth-century bourgeois consumption was important and multipurpose: it "served to define their family's history, to signal their social position to other bourgeois, to differentiate their class from that of aristocrats and workers, and to lay claim to the nation's grand past."[10] Even though the mistress did not herself prepare the artistic pieces, she expressed her good taste and degree of refinement in the appropriateness of her choices as a consumer, which in turn indicated the status of the entire household.

For the elite in particular, membership in a selective taste culture determined one's reputation; deploying sugar sculptures at parties and banquets was one way they articulated that membership. Before the influx of European immigrants and the subsequent influence of German traditions in particular, for example, American confectioners drew on classical forms for reference and inspiration. In 1844, Parkinson's *Complete Confectioner* instructed that the successful "artist" working in sugar "must be guided by his own genius and fancy." But for ideas, Parkinson recommended consulting James Page's *Guide for Drawing the Acanthus* (1840), which described and illustrated in fine detail the various compositions made from the acanthus leaf, the basis for all neoclassicism.[11] Not only was the acanthus leaf a versatile design motif that could give rise to many others (making it a perfect source of inspiration for the burgeoning confectioner), but it also, as the quintessential element of neoclassical design, symbolized the republican values of virtue and stability.[12] A visual example of the neoclassical influence appeared as an engraving in the December 1858 issue of *Godey's Lady's Book,* entitled "Christmas for the Rich." A typical depiction of the well-appointed nineteenth-century family gathered around the glowing hearth, the scene shows members enjoying one another's company and their Christmas bounty. Completing the iconography of the scene, a black servant enters the room carrying an elaborate sugar sculpture—a spired dome—on a tray.

Herman Geilfuss, a German immigrant working in Philadelphia, made sugar sculptures like the one illustrated in *Godey's.* Calling himself a "Manufacturer of Ornamental Confectionery" since opening his Vine Street establishment in 1862, he enjoyed a successful and loyal trade with patrons: private customers, hotel operators, and other professional bakers called on his abilities as a fancy sugar worker. Men like Geilfuss stood apart from the more common German bakers who baked loaves of bread in often illegal and unsanitary basement bakeries. Geilfuss, for one, profited from his talents as an artisan and enjoyed a thriving business into the 1880s, when a Philadelphia guidebook wrote, "He has from a comparatively small beginning built up a large and lucrative trade, occupying the entire building, and employing six to eight assistants. His manufacture is of a very superior character, and he has made, and is constantly making many improvements therein."[13] Geilfuss possessed the abilities and versatility typical of the nineteenth-century sugar worker. He not only accommodated custom orders by executing myriad designs in sugar but capitalized on the growing popularity and commercialization of holidays by selling articles in specific celebratory motifs, including wedding cake ornaments, Easter eggs, and Christmas confections.

In many instances, people developed their tastes from what they saw at the local confectioner's or what they read in printed accounts. For example, *Frank Leslie's Illustrated* often relayed news about banquets across the United States and included both textual and visual descriptions of such events. Reporting on the 1856 Railroad Reunion Ball in Buffalo, New York, the author stood in awe of the sugar sculptures. One, a "very excellent model" depicted the

"Portage Bridge and Falls" by rendering a locomotive and cars in sugar along with the "Falls"—most likely in strands of spun sugar—cascading down banks made of "miniature rocks in sugar." On the same banquet table and measuring eight feet tall stood "a huge pyramid in crimson and white confectionary, with gothic windows on every side." Finally, on the same table was "an immense sugar vase in blue and white, which was almost a perfect imitation of Parian marble."[14] Other fêtes also extensively covered in the press were those hosted by Mary Todd Lincoln. Accounts meticulously described the kinds of ices, ice creams, ice sculptures, and sugar sculptures supplied for her lavish parties. One such, the Grand Presidential Party, held in Washington on the night of 6 February 1862, and catered by Maillard, was the "exclusive topic of the beau monde of Washington."[15] Worthy of its corps of distinguished guests, the food itself presented a "coup d'oeil of dazzling splendor, fruits and flowers, and blazing lights, and sparkling crystal, and inviting confections."[16] The *New York Herald,* a popular newspaper that disseminated information about the activities of the upper classes, deliciously related the event and its trappings, listing in detail, among other treats, the "candy ornaments" present at the occasion:

> A representation of a United States steam frigate of forty guns, with all sails set, and the flag of the Union flying at the main.
>
> A representation of the Hermitage.
>
> A warrior's helmet, supported by cupids.
>
> A Chinese Pagoda.
>
> Double cornucopias, resting upon a shell, supported by mermaids, and surmounted by a crystal star.
>
> A rustic pavilion.
>
> The Goddess of Liberty, elevated above a simple but elegant shrine, within which was a life like fountain of water.
>
> A magnificent candelabra, surmounted by an elegant vase of flowers and surrounded by tropical fruits and birds, tastefully arranged and sustained by kneeling cupids, holding in their hands a chain of flower wreaths.
>
> A fountain of four consecutive bowls, supported by water nymphs—an elegant composition of nougat Parisien.
>
> A beautiful basket, laden with flowers and fruits, mounted upon a pedestal supported by swans.[17]

Fancy sugar pyramids, whether imitating a rustic pavilion or a tiered fountain supported by water nymphs, made stunning visual impressions, worthy of the occasion at which they appeared, and conversely marked the occasion itself and its subtleties as equal partners in this high culture. Mrs. Lincoln, as one tastemaker, subscribed to a centuries-old tradition that considered such fancy ornamental confections impressive and entertaining, whimsical in subject yet serious in class connotation. Individuals who wanted and were able to purchase such pieces needed to look no further than the local skilled confectioner, whose ads—often illustrated with representative work—appeared in newspapers and periodicals.

The Confectioners

The culinary contributions of immigrants to the United States plus the growing availability of refined sugar, their medium, led to the terrific profusion of stylistic variations common by the end of the century, from *pièces montées,* or set pieces, sugary replicas of recognizable objects, to ornate piping on ceremonial cakes. Many in Europe had enjoyed fancy sugar goods for centuries, and French, English, Italians, and Germans all practiced various forms of the art. In the United States, the Germans dominated the field of ornamental sugar work, continuing not only their own aesthetic traditions but also a form of the guild system that systematically trained these fine confectioners.

To their advantage, the Germans also organized and established a number of professional associations, including the Journeymen Bakers National Union of the United States. In addition to both local and national professional groups, bakers' trade literature also fostered a sense of German identity, solidarity, and specialization of vocational knowledge. For example, the *Confectioners' Journal,* which began publication in 1875, featured recipes written in German. The *Deutsche-Amerikanische Bäckerzeitung,* yet another trade journal, began using the two languages in 1895, becoming the *Bakers' Journal and Deutsch-Amerikanische Bäckerzeitung.*[18] Each issue of the *Bakers Weekly,* first published in 1904 in New York City and written for the members of the Master Bakers, was divided in two—half written in English and the other in its German translation. In addition, the *Deutsch-Amerikanisches Illustrirtes* [i.e., *Illustriertes*] *Kochbuch* (*German-American Illustrated Cookbook*), directed at professional chefs and containing over 1,000 recipes for highly ornamental dishes,

FIGURE 27. An example of German skilled sugar work from an instructional book. "Füllhorn von Mendeln mit caramelierten Früchten," in Charles Hellstern, *Deutsch-Amerikanisches Illustrirtes* [i.e., *Illustriertes*] *Kochbuch* (New York: G. Heerbrandt, 1888), 420. Courtesy, The Winterthur Library: Printed Book and Periodical Collection.

was published in New York in 1888. Written solely in German, this book emphasized ornamental work and included detailed instructions for sugar sculptures and other table decorations (fig. 27). Social, professional, and published institutions defined and bolstered the occupations of bakers and sugar workers and provided instructions on how to succeed in them.

Trade literature often encouraged ordinary bakers to add ornamental sugar work to their repertoires. John Hounihan, in his 1877 *Bakers' and Confectioners' Guide,* distinguished between regular bakers who baked only bread and "cake bakers," who, more skilled, had a greater earning potential: "Any baker can make cakes, but to make a fine cake baker, a man must give his whole attention to it." Unlike ordinary bakers, fine cake bakers decorated what they baked, exercising "great care and neatness, as cakes are not a necessary thing of life, and makers must work to attract the eye and suit the taste." Hounihan extolled cake baking because it was more lucrative and less onerous than bread baking. In 1877, a successful cake maker earned from $10 to $15 a week, while a bread baker working over twelve hours a day was lucky to earn two-thirds of that. Rather than chiseling marble or modeling clay, the professional cake maker pulled, spun, baked, molded, and colored refined sugar to create forms architectural and figurative, monumental and personal, abstract and realistic. "The fine arts are five in number: Painting, Music, Poetry, Sculpture, and Architecture, whereof the principal branch is Confectionery," the famous French chef Marie-Antoine Carême pronounced.[19]

Ornamental sugar work remained a male-dominated profession throughout the nineteenth century, although there is little indication that sugar workers in America directly trained others in the same art. Instead, men who showed an interest relied on the trade literature and books like J. Thompson Gill's *The Complete Practical Ornamenter* (1891), which offered to teach the confectioner and baker "a practical knowledge of the processes of Ornamentation. . . . which will enable him to imitate, in sugar and other edible articles, any scene, picture, or any event taken from art or nature, or which may be the invention of fancy."[20] Published works devoted to the bakery and confectionery professions expressed active interest in teaching readers how to produce successful, artistic ornamental sugar work by including step-by-step instructions and diagrams. But they highlighted an established sugar worker's repertoire as much as they instructed novices, and articulated the chasm between works that a beginner could attempt and those achieved only by the master confectioner, who belonged to a rare breed. This hierarchy conferred on sugar sculptures a double status: they were the unique and time-consuming products of a talented confectioner's artistic vision, hence they were expensive. In addition, well-known confectioners acquired reputations, which also added to a piece's social prestige and market worth.

Most sugar sculptures that appeared in regular shopwindows, made by agile hands and clever minds, still could not match works generated by the pantheon of fine confectioners who enjoyed star status in the pages of the trade literature and among their wealthy patrons, similar to today's cult of the celebrity chef. Month after month, turn-of-the-century trade journals, which featured the work of famous ornamental confectioners, glamorized the work and presence of these talented men. Column upon column pictured them posed alongside (or instead of) their work and described who they were and for which famous American and European hotels and restaurants they worked. Those who possessed and promoted their more specialized skills beyond basic bread baking could do very well. For example, the *Bakers Review* regularly featured the work of the famous New York confectioner Philip Goetz, even claiming that his name, by 1903, had become a "household word."

Representational miniatures in sugar combined capitalist ideals of privilege with ideas of entertainment, conforming to prevailing cultural ideologies. Sugar sculptures, nineteenth-century versions of medieval subtleties, did not seem out of place to Victorian Americans but actually tapped into their hopes and fears of the time in a material way. The late nineteenth-century American bourgeoisie's interest in medievalism reflected changing beliefs about the maturation of American culture based on social Darwinism, T. J. Jackson Lears observes. "Childish traits characterized any society which did not conform to the model of Western industrial capitalism. The 'civilized world,' having put childish things in its medieval past, had reached social adulthood." Such childlike qualities were nonetheless also admired, because they triggered feelings of innocence and spontaneity lacking in restrained Gilded Age society. Sugar sculptures were objects that the bourgeoisie could possess, the way a child possessed a dollhouse, but in a refined and knowing way. They also served a function similar to that of pleasure gardens, facilitating the "deliberate cultivation of spontaneity . . . temporarily recapturing the outlook of children, medieval knights, and contemporary primitives."[21]

Common at banquet tables, sugar displays also found their way to consumers through the paths of commercialism itself. The shopwindow, already a distinct theatrical space, became even more theatrical when furnished with a piece of figurative sugar work. Miniatures taken "from nature, from works of art, from suitable pictures, or from the scenes that are daily before" one, as one confectioner advised,[22] these pieces became both toylike and alive—yet another form of sophisticated plaything for elite adults, their expensive penny candies. Although plate glass windows prevented shoppers from interacting with these objects directly, they could activate fantasy worlds just through sight alone.[23] A Philadelphia guidebook from 1887 described how shop owners, in particular "*elite* confectioners," used display to draw people in by featuring one lone item "that the beholder may, as *ex pede Herculeum,* divine from this very slight indication that something very superior is to be had within."[24] And this visual aspect was so important that trade journals repeatedly implored confectioners to decorate their windows with "eye catching" displays that would attract attention and draw in business. Most of these journals also featured monthly columns suggesting new ideas for window decorations.

Whether seen as monumental pieces of artistry or whimsical novelties, sugar sculptures evoked both admiration and fantasy. *Frank Leslie's Illustrated Weekly* described New York City streets before Christmas in 1872 as exhibiting "the most practical democracy," with rich and poor alike thronging around shopwindows, and commented: "Passing up Broadway with this mixed company, stopping here and there to notice the strange contrasts that were brought to view by every display of holiday goods, we found a large and eager party before the windows of Purssell's *confiserie,* wondering at the manipulation of candy and fancy cakes for New Year's tables" (fig. 28).[25]

Confectioners, contemporary sorcerers, transformed diverse materials into magical scenes. A "rock work" windmill, for instance, emerged from rock candy, icing, sugar cubes, spun sugar, colored granulated sugar, and dried bread. Nonedible elements layered illusion upon illusion and included tinfoil (to imitate water), flake glass (snow crystals), artificial moss, papier-mâché, and colorful dyes. At times, sculptors even used nonsugar materials to imitate sugar that was supposed to simulate something else. For example, rather than using real foliage to create the trees in the scene surrounding a "Village Church," the author suggested starting with a twig that would provide the best miniature treelike shape and then coating it with layers of sugar: "The trees may be made by picking small twigs, dip them into thin icing and draining it off. When dry wash them with gum water and dust on sugar. . . . If you wish heavy foliage wash these parts two or three times, then add sugar, after each washing and drying."[26] An essential part of the enchantment of the work was that it was made out of sugar; sugarcoating nonsugar elements was just one way to achieve and maintain the artifice while keeping the sugar component foremost in people's minds. Even by the end of the century, when sugar was cheap and plentiful, this conceit still managed to conjure the lifestyles of past and present European royalty, a perennial interest of American popular culture.

Like their medieval precursors, Gilded Age sugar sculptures blurred the

FIGURE 28. Confectioners at work could be as spectacular as their creations. "Life Sketches in the Metropolis," *Frank Leslie's Illustrated Newspaper,* 13 January 1872, 284. Library Company of Philadelphia.

distinctions between food and art, the functional and nonfunctional. Even so, many confectioners' advice columns warned against using too many bright vibrant colors, because these pieces were judged as confectionery, and their ultimate value *as* objects made of sugar depended on their being "refined." Garish displays revealed a confectioner's lack of taste.[27] Inedible ornaments, wrote one author, "may be made of plaster, gum paste, wax, wood, tin, glass or cardboard, and a little more latitude in the matter of colors is admissible, still it is not wise to indulge too freely in color, even in these—harmony in color counts for more than gaudy contrasts—and the general public are opposed to loud colorings."[28]

The Eiffel Tower, castles, laboring gnomes, ocean liners, and windmills proved equally worthy subjects for the confectioner. Some confectioners even found inspiration in products of industry, replicating contemporary mass-produced objects such as automobiles and table lamps in sugar. Regardless of subject matter, these pieces were both works of art and demonstrations of skill.

⁖ *Wedding Cakes* ⁖

Wedding cakes, the most popular and prevalent genre of ornamental confectionery, combined figurative elements and a sense of the monumental. The iconic wedding cake—white, three-tiered, and decorated with frosting flowers—first appeared in the late 1870s. In material form, wedding cakes simultaneously encapsulated many competing consumer preferences. Couples wanted something tasteful and controlled, which the confectioner articulated by drawing on abstract, geometric forms. Yet they also wanted something monumental, to present as a tangible proclamation of marriage. In addition, they wanted something considered a unique work of art, evincing the sensibilities of the bride and groom and serving as a memorable status symbol. From the Gilded Age on, couples increasingly opted for formal wedding ceremonies, and the wedding cake filled a gap in the ritual. Drawing on English, German, and Italian traditions, and catering to many tastes and budgets, confectioners produced wedding cakes of many kinds, from the singular, unique work commissioned from a famous sugar worker to humble assemblages of ready-made elements. For the comfortable classes, the wedding cake marked the union itself and indicated social status, representing upper-class aspirations and accepted cultural mores.

The late nineteenth-century wedding cake had moved worlds and centuries away from its precursors. According to English tradition, ceremonial crackers made of wheat, symbolizing abundance and plenty, were broken over the wedded couple's heads as they crossed the threshold of their new abode. Everyone present shared in eating this communal nod to future success. Breaking the crackers over the couple's heads, the community overtly sanctioned the union and recognized it as beneficial to the welfare of the whole society.

After the French Revolution, fancy confectioners exiled from their homeland and in need of clientele emigrated to England. There they replaced the elite's traditional wedding biscuits with stacks of wedding buns covered with a thick icing, thus launching the quintessential pyramid form of the wedding cake that we recognize today. Not long afterward, confectioners decorated these tiers of baked goods with subtleties, "figurative of the joys of matrimony,"[29] which added to the cost of the pyramids and rendered each one unique. Wedding cakes, as simple baked goods, represented communal acceptance of a union and the incorporation of the couple as a productive unit into the larger society. Professional confectioners turned them into showpieces that, rather than celebrating group cohesion, represented the atomistic union of the couple. "Cakes for breaking belong clearly in the earlier world in which they were playing a part in a palpable rite of passage for the bride," Simon Charsley writes. "This was something done to her by others, an assertion if you will of the far from private transition she was making."[30]

In the eighteenth century, the English preferred dense, dark, one-tiered wedding cakes laden (like our modern conception of a fruit cake) with fruit and nuts, or mincemeat. An outer layer of decorative icing established the cake's significance. According to Simon Charsley, the American wedding cake, unknown before the late 1700s, was based on this English model until it was modified in the later nineteenth century. Yet even early in the century, the tradition of the wedding cake was already so familiar that a writer in an 1831 issue of *Godey's Lady's Book* likened it to the marital relationship itself, sardonically suggesting that the sham decoration concealing the cake's underlying ingredients was analogous to the couple's similar subterfuges: "The sugar was only a covering to the carbonized surface, the eating of which discovered itself in the honied terms of 'my love,' and 'my dear,' that are first all sweetness, but soon discover the crusty humour beneath."[31]

English royalty elevated the wedding cake, making it three tiers and sometimes five feet high, a form quickly appropriated by fashion-conscious Americans. The Rensellaer wedding cake of 1853, for example, typified the old-style cake adopted by America's elite, which was much more reminiscent of a medieval subtlety or John William Millers's eighteenth-century Philadelphia creation than a modern cake. Over six hundred guests attended the wedding; yet the focal point proved not to be the marrying couple but the bride cake, a "beautiful, *à la Renaissance* production," described as follows:

> Upon the octagon platform, supported by eight Cupids holding garlands of flowers, is represented a classic marriage procession; mounted upon a car of exquisite workmanship, drawn by doves, comes the "god

> Cupid," clasping to his breast his bride elect, the lovely Psyche. The car is preceded by musicians and pages, scattering flowers upon the road; torchbearers, and marshals, with their staves of office, to maintain order throughout the line; on the opposite side is a triumphal arch, with little Cupids entwined among the same. It displays a motto of "Vive l'Amour," and through the arch is seen an altar, with priests in attendance, whose sacred fire announce to the gods the consummation of the holy act; in the center of the platform stands a magnificent vase, upon a small temple,—the whole cake, an establishment, is supported by ornamental scrolls and pedestals.[32]

Newer versions of older culinary status symbols, cakes such as these served a dual purpose—to mark the significance of the wedding ceremony and to confer luxury on and therefore define the importance of the wedding party and guests. Constructed by professional men particularly skilled in sugar paste work and fine confectionery, these cake sculptures proclaimed their import by their very monumentality and continued the English tradition of marking an important event with a special cake. "God-cakes" ushered in the New Year, other cakes celebrated the Feast of the Epiphany, and pancakes indicated the approach of Lent.[33] But until the 1860s, these cakes remained primarily English affairs, largely disregarded by Americans. The American angel food cake, a whiter, lighter cake, eventually became the preferred alternative to the British fruitcake, and was called the "bride's cake." Whatever the internal composition, the outward appearance of the British versions influenced what the American wedding cake eventually looked like.

Except for the birthday cake, which became popular after the turn of the century, the late nineteenth-century wedding cake was the only ceremonial cake to gain widespread acceptance in America. It had deep roots encompassing many ethnic traditions, yet its basic American form also allowed for variations based on one's own taste and economic ability. One contemporary author, writing in 1900, noted that the wedding cake "has as much care taken in its construction as if it were intended to stand a lasting memorial of marital felicity, instead of being the perishable creation of a nuptial day." Cakes rendered in the German tradition tended to be single-tiered and shaped like cornucopias, referring to fertility and abundance. But English-influenced versions reigned: three-tiered and highly decorative affairs, they signified status and power, jutting into the air like skyscrapers. In addition to their symbolic qualities, wedding cakes at this time also directly expressed the taste of the wedding party, and hence indicated aesthetically determined and articulated class distinctions, so that "what Sal and her steady describe as 'real elegant' may not seem beautiful to those who move in a different circle and whose esthetic canons are not the same."[34]

As a singular event in a person's life, the wedding ceremony incorporated some amount of ostentation even for people who were not rich. It is not surprising, then, that Gilded Age Americans, an emulating sort, chose the wedding cake as the object on which they lavished expense and attention, especially since confections had propelled the courtship process to this final point. The culmination of that process needed a confection equal to the significance of the event, and also one clearly symbolizing the political meanings of such a rite. Indeed, women who first saw their marriage ceremonies outfitted with such spectacular cakes were among the first generation of Americans to have known democratized sugar throughout their entire lives. When wooing a woman, a man's choice of chocolates—everything from the shapes and fillings to the packaging and brand—conveyed his taste, culture, and income, in addition to underscoring his role as provider of a woman's pleasure. The wedding ceremony in general and the cake in particular served as logical (and terminal) extensions of this process. Soft candies were to the fiancée (*as* fiancée) as wedding cakes were to the bride *as* the bride—the two were conflated, inseparable, synonymous.

Concern about the latest fashions applied equally to the cake and the bride. Although some kind of ceremonial cake had accompanied well-to-do American weddings since the eighteenth century, and an 1868 American etiquette book referred specifically to a "wedding cake," the monumental ceremonial piece that we know of today as the archetypal wedding cake really caught on in America in the early 1870s.[35] Princess Louise's 1871 nuptial ceremony featured a huge cake made by Queen Victoria's chief confectioner at Windsor Castle. Standing over five feet high and measuring two and a half feet in diameter, this "miniature temple" was embellished with the "allegorical figures of Agriculture, Fine Arts, Science and Commerce" and topped with a "Vestal Virgin," reminiscent, once again, of medieval subtleties (fig. 29).[36] One writer described the cake after the fact as "purely classic in design, each tier rising on Pillars, and adorned with exquisitely molded statuary."[37] When ac-

FIGURE 29. Princess Louise's wedding cake sparked the American craze for similar objects. "The Royal Wedding Cake," *Illustrated London News,* 1 April 1871, 308. Library Company of Philadelphia.

counts of Louise's cake reached America, they inspired a new trend in weddings, which suddenly required enormous, elaborate wedding cakes. In the last decades of the nineteenth century, wedding cakes for the rich became increasingly ostentatious and patterned after those of British royalty.

The reputations of the confectioners making these cakes depended not only on the skills they demonstrated but also on the public's perception and approval of their aesthetic decisions. Although, as hired workers, they had to meet the needs, wants, and tastes of their clients, they did not always have to approve of them. For example, an article appearing in a 1901 issue of the *Confectioner and Baker* detailed many British wedding cakes that fell outside American ideas of proper taste. The author described the "rather appalling" cake made for the duke and duchess of Cornwall and York for their wedding, two of whose tiers "alone weighed together between four and five hundred weight, and had a combined height greater than that of the bride and bridegroom." In another contemporary example, a "New York undertaker recently, with deplorable taste, insisted on having his bridal cake decorated with skulls and cross-bones and other grim suggestions of his profession."[38] Although cultural arbiters looked upon this kind of iconography with disdain (for it went against even the minimal definitions for what constituted proper cake decoration and proper fashion), it is notable for its very idiosyncratic nature: upper-class men with enough money could have something completely personalized, sometimes going beyond the bounds of acceptable standards for a wedding cake. The cakes of the middling classes tended to incorporate more universally acceptable design elements, classic Greco-Roman style, which endured and were enduring, extravagant yet restrained.

The more common American wedding cake, which remains popular today, was a primarily white, three-tiered arrangement decorated with icing flowers and latticework piping, topped with overtly symbolic cake ornaments. By the last two decades of the nineteenth century, the wedding cake had come to closely resemble the bride herself, who wore a white dress punctuated with flowers (usually orange blossoms), and a veil, tiara, or other headpiece (fig. 30). "What we think of today as the wedding cake is actually the 'bride's cake'—frilly, decorative, not meant to be eaten. The 'groom's cake' was dark, a fruitcake, practical, substantial. The sexist implications rise like yeast," Marcia Seligson comments.[39] Indeed, some say the use of flowers on wedding cakes

FIGURE 30. Popular fashions simultaneously influenced bride and cake, giving them the same aesthetic sensibilities. "Godey's Fashions for December, 1877," *Godey's Lady's Book,* December 1877. Library Company of Philadelphia.

(fig. 31) derives from the centuries-old use of "favors," fabric bows lightly sewn to a bride's dress. Immediately after the ceremony, the guests descended upon the bride, ripping off her favors to take home for themselves.[40] The bride's cake, wrapped in floral garlands, symbolized the betrothed, pieces of which as body parts became souvenirs to be shared among the guests. Cutting the cake literally and figuratively distributed these "favors" of the bride, but obliquely enough to suit Victorians' strict sense of propriety.

That the wedding cake and the bride, both of which could have looked much different, arrived at a similar appearance by the end of the nineteenth century was not accidental. They were both ornamental showpieces that confirmed the status of an entire family, and established the dominance of a husband over his wife by equating her with a decoration whose role and appearance he could determine. Thorstein Veblen, again, is instructive: "[A woman's] sphere is within the household, which she should 'beautify,' and of which she

FIGURE 31. "Design for Bride Cake," *Bakers Review*, 15 January 1903, 29. Library of Congress.

should be the 'chief ornament.' The male head of the household is not currently spoken of as its ornament."[41] Icing that decorated a cake, much like lace that embellished a dress, held more iconic than actual value, and so was simultaneously superfluous and essential. In fact, the kind of sugar icing required to frost a cake did not even taste good. As one trade journal commented, "In the case of the bride cakes, the sugar is used more for ornamental purposes than for the fact that it is a pleasant or palatable article of consumption. . . . bride cake icing is hard and unpleasant to eat."[42] In contrast, the dark and edifying groom's cake, which was, eventually, completely supplanted by the white bride's cake, could sustain the guests. It was substantive and had nutritional value, while the bride's cake stood to be appreciated for its beauty alone.

Like medieval subtleties, pieces of which accompanied distinguished guests home, slices of wedding cake packed into tiny boxes also left upper-class ceremonies with guests. "Upon departing," wrote the *Boston Cooking School Magazine,* "each guest supplies himself with one of these souvenirs. Occasionally a maid is stationed in the hall to hand the boxes to guests as they pass out."[43] Many confectioners provided pre-cut and pre-frosted pieces of cake in small boxes—often embellished with the couple's initials in gilt—tied in satin ribbons to be distributed to guests at the wedding or mailed to those who could not attend the festivities. A verse published in *Selections for Autograph and Writing Albums* of 1879 showed the popularity of this practice even before the turn of the century:

Remember me when far away,
  If only half awake;
Remember me on your wedding day,
  And send me a piece of cake.[44]

Pieces of wedding cake, fragments of the larger cake itself, were small and consumable representations of the ceremony and the bride herself—both of which guests otherwise partook of, necessarily, only as spectators.

Because the bride's cake was mainly decorative, slices of dark fruit cake, distributed in boxes, served as surrogates, "prepared by the caterer, one for every guest, and are meant to serve in the place of the slice from the bride's loaf to which, in other days, every guest was entitled."[45] In either case, each person received one piece, and, "To take more than one box, unless asked by the hostess to convey one to a friend," warned one advisor, "is unpleasant evidence of very bad breeding," and perhaps suggested an indelicate amount of ardor for the bride, if, indeed, the cake represented her body.[46] The top of the cake, often fabricated from nonedible sugarcoated papier-mâché, wire, and cloth, also functioned as a personal keepsake for the bride and groom, which they removed from the cake before cutting: "The top ornament on the board can be kept as a souvenir, or made to do duty a second or third time."[47] People tended to keep these ornaments for a long time and even gave special consideration to pieces of cake taken home from the ceremony. Families often treated such pieces of cake like relics, placing them in tiny wooden boxes lined with tissue paper for posterity. Solidified (and well-preserved because of their high sugar content), they passed down through subsequent generations, who carefully labeled them and treated them as special keepsakes.[48]

By the end of the nineteenth century, a wedding cake accompanied almost every marital ceremony, even though many people complained that, with so much attention paid to ornament, the white cake lacked taste, and the frosting enhanced only its appearance and not its palatability. One advice book of 1894 made eating wedding cake a duty: "No matter what one's private convictions regarding the qualities of wedding cake as an edible may be, it is expected that each person will take a piece and eat, at least a crumb."[49] The otherwise superfluous decorations that frequently rendered the wedding cake inedible also endowed it with the symbolism that gave it its purpose.

The bride's cake symbolically reiterated the bride and alluded to her function in both the immediate wedding ceremony and the larger context of her future social and familial roles. Because the cake and the bride were as one, what was done to the cake physically was done to the bride metaphorically. Therefore, gestures like cutting the cake became highly significant. "Plunging the knife into the centre of the cake breaks through the 'virginal white' outer shell,"[50] inverting the long-past tradition that directed wedding guests to break a biscuit over the couple's heads in a display of community solidarity. Since the wedded couple now cut the cake themselves, the act represented the breaking of the bride's hymen by her husband, the consummation of the marriage, and the larger power dynamic between the wedded couple. In addition, many accounts referred to the bride cake as *his*—meaning, belonging to the groom, and explained that the groom decided how it was to be decorated, often with ornaments relating to his interests or profession. The bride and cake were thus equally possessions of the groom and meant for his prestige.

Brides and grooms relied on the expertise and artistry of the confectioner to convey just what they wanted: pieces of custom-made art, wedding cakes were very expensive. "In the hands of a really expert confectioner the wedding-cake becomes a veritable work of art, an ornate structure into which it would seem vandalism to stick a knife."[51] Hence the reason often for two cakes—one for eating and one for show. The author of a short story about a turn-of-the-century wedding noted this duplication, saying of the reception, "There was a beautiful, great monument of a cake with a ring in it. . . . (Of course the real wedding-cake was in richly decorated white-satin boxes in the hall for each guest to take at the moment of departure.)"[52] Meant to please the eye and symbolize the moment, fancy wedding cakes sometimes cost over $300. A caterer who supplied one such cake accounted for the great cost of some of these pieces, saying that "nowadays a wedding cake, to be a wedding cake, must be a creation by itself. In the first place, each fine cake must be flavored and fruited after a new fashion. This necessitated the exercise of much ingenuity and skill on the part of the chef. Then the design had to be specially made by the sugar cook." He continued: "A month ago a wealthy young New Yorker, who is a great breeder of dogs, got married. His bridal cake had a portrait, in wax and colors, of his favorite St. Bernard on top, and by its side his bride's favorite collie."[53] The total weight and amount of decoration determined the cost of cakes requiring less customization. Because the dogs had to be specially made by a professional sculptor, the cake cost $300. One professional scale, from 1904, calculated that a cake by "the best" confectioners cost from 40 to 50 cents a pound without decoration. Ornamentation added another $2 to $10, depending on the size of the cake and "class" of ornamentation. More generic cakes, "iced piped, and just ordinarily decorated," cost from 50 to 60 cents a pound, with everything included.[54]

⸫ *Mass Production* ⸫

Economic considerations forced most turn-of-the-century betrotheds to have only one cake, and being such a powerful symbolic form, the bride's cake endured. Technological innovations led to affordable cake decoration, and machine production affected even fine confectionery, historically defined by its handcraftsmanship. Ironically, the same technology that popularized and democratized the products of fine confectioners also jeopardized their livelihoods by threatening to replace most handwork with mass-produced goods.

Confectionery numbered among many trades, including plasterworking, woodworking, and iron casting, improbably replaced by the efforts of machines.[55] In his 1891 treatise on confectionery, J. Thompson Gill suggested that "those who wish to find additional material for invention or imitation, [should] consult the catalogues of designs for scroll-sawing, the designs in certain fashion journals, or the works of architecture, and illustrated magazines."[56] Over fifty years earlier, Parkinson advocated looking at nature and consulting works showing examples of the classic acanthus leaf. Now, confectioners were to seek out industrial designs for their inspiration. Not only had these forms of workmanship become more accessible, but people marveled at the ability of machines to impart such complex shapes onto and into things like wood, iron, and even sugar.

In ornamental confectionery, as in other crafts, technological advances uneasily coexisted with handwork for quite a while.[57] For example, trade journals regularly featured detailed instructions for decorating fancy cakes by hand, while at the same time carrying advertisements for mass-produced ornamental goods. The technological advancements used in the manufacture of decorated cakes both aided and crippled the work of the sugar craftsman, who spent years learning his occupation and promoting the merits of his ornamental confectionery. Confectioners experienced the classic paradox of industrialization: mass-production democratized formerly fancy goods, making them accessible to more groups of people. At the same time, however, it drove into obsolescence the very handwork that had inspired and given rise to such products in the first place.

The ability to lay down fine, even lines of sugar fondant piping, to create intricate sugar latticework, to form, paint, and position marzipan shapes, even to frost a cake with icing of uniform surfaces and matching whiteness, all separated the products of the expert sugar worker from crude imitations. Machinery and the specialization of labor, however, no longer confined these skills to a discrete trade. The individual confectioner now found himself competing not only with women who were making and decorating cakes in their homes, from recipes in domestic manuals and ladies' magazines, but also with huge confectionery supply houses that sold ready-made sugar work at cheaper prices. The supply houses could afford to do this because they used assembly-line methods to produce standardized pieces, which were then put together to create "individualized" decorations.

Instructional manuals, which continued to praise the original work of the confectioner-as-artist, largely decried these ready-made figures and cake elements. The use of gum paste, a mixture of gum Arabic, glucose, and flavoring, to make cake ornaments was, for example, considered "somewhat a matter of regret" by the authors of *The Victorian Book of Cakes* (1902), which observed that "most of this work is done by the mechanical means of molding, which, unless the workman should happen to be a skilled modeler, would mean the use of other men's designs and workmanship."[58] But in order to stay competitive, many bakers and confectioners relied on ready-made embellishments to decorate cakes. Schall & Co., for one, manufactured and imported all kinds of ornamental goods for the modern confectioner (fig. 32). While they also sold all manner of "holiday novelties," lace papers, and almond pastes, they specialized in ready-made wedding cake tops constructed from any combination of sugar, marzipan paste, and papier-mâché. Similarly, the Bruce & West Manufacturing Co., of Cleveland, Ohio, also manufactured fancy ornaments. Their early twentieth-century trade catalogue listed over 270 different wedding cake ornaments. The designs ranged from the simple to the elaborate, reflecting the wide variety of consumer tastes and pocketbooks. Their sheer number also indicated just how popular wedding cakes had become in the past thirty years. Bruce & West's cheapest designs (sold wholesale for 20 cents each), consisted of modest ornaments whose dimensions were too small to note in the book. They included, "number 48, Flower, with No. 2 Dove and Letter"; "number 49, Rose, medium, with No. 1 Dove and Letter"; and "number 50, Rose, large, with No. 1 Dove, Letter and Leaves." Mass-produced varieties of cake ornament design merely appropriated standard Victorian forms and common sentimental tropes. Requiring no unique skills to assemble and warranting no special aesthetic or manual attention, these ornaments were affordable to the poorer classes, who could finally mark special occasions in a middle-class way without bankrupting themselves.

The middling sorts more likely preferred "number 116, Cupid in Shell and Swan with Flowers," which at 10 inches high and 6½ inches in diameter, sold wholesale for a dollar. For the people who wanted to go all out but who still could not afford the customized work of the skilled confectioner, Bruce &

FIGURE 32. Ready-made cake ornaments helped democratize the fashionable wedding cake. Schall & Co. broadside, ca. 1893. Collection of the author.

West offered number 152, their most expensive piece, which their catalogue described as "Four Horns of Plenty, large Flower Basket and Cupid with Silver Bell," which rose to 23 inches, had a diameter of 11 inches, sold wholesale for $7.15, and must have been intended for a gigantic cake. Clearly, the profusion of these ornaments in various styles and prices reflected a company's willingness to consider and cater to consumers with many different tastes and budgets. One could top his wedding cake with "Dwarf pushing Wheelbarrow with Flowers," a "Large Wish Bone with Doves and Spray," a "Sailor Boy holding Anchor under Silver Bell and Flowers," or, ominously, a "Large Cupid crying Broken Heart, Dove and Spray."

These examples represented the collision of fine confectionery and mass production, which occurred at the end of the nineteenth century. Enterprising confectioners opened up their own manufactories in order to capitalize on the burgeoning trade in machine-made turn-of-the-century wedding cakes. Simon Charsley notes that the popularization of such ready-made cake decorations reflected not only technological encroachment but also a shift away from the "purist" ornamentation of piping and frosting, which could only be obtained from a skilled confectioner: "On the one hand, neither the purse of the customer nor the skills of many confectioners could rise immediately to the fulfillment of the ideal, on the other, there was always some demand for decoration which would be more directly related to the occasion."[59] Piping and other icing ornamentation could not be reproduced by machine or through rationalized bits of hand labor; only molded forms, such as marzipan and sugarpaste figures could be so manufactured. Ready-made ornaments meant greater variation among the confectioners and customers alike, who chose based on a confectioner's skills, a decoration's affordability, and one's own desire to "personalize" the occasion (perhaps accounting for eccentric combinations such as "Bride and Groom with Bicycle, Marriage Bell, Good Luck and Flowers"). The people who selected such ornaments for their wedding cakes literally and figuratively bought into the "good life" as proffered by the elite. While the ultra-rich hosted ceremonies replete with customized cakes that expressed the occasion and their individual personalities (recall the couple's cake festooned with miniature replicas of their hunting dogs), the middling now enjoyed quasi-customized cakes that were the "unique" results of mass production.

Not surprisingly, at the same time as this profusion of ready-made ornamentation became available, tastemakers decried the symbols and images produced in such a manner, shifting their markers of high culture once again. They shunned Victorian ornament and only regarded as "pure" neoclassical or baroque forms calling for fine, expensive, skilled handwork. The elite looked down upon more "readable" and less abstract figurative forms—the common allegorical images like flowers, cupids, doves, and hands that made up the elementary vocabulary of Victorian sentimentality.

Members of the confectionery profession themselves expressed mixed feelings about the shifting tastes and production methods found in ornamental confectionery. Gill, the expert cake ornamenter and sugar worker, expressed many artisans' concerns at the time in his late nineteenth-century instructional book. Most notably, he wrote of mass production's encroachment on their skilled labor, stressing the diligence required of the novice who was just learning about skilled sugar work and warning against thinking that "the art of Ornamental Confectionery is one that is easy to be mastered." He continued, "It is true that the novice can produce many striking and good effects by the exercise of a little ingenuity, and by using the vast amount of gum paste ornaments, artificial foliage, etc., which can be had of any of our supply houses; but to become an artist in sugar and kindred materials requires both study and perseverance."[60] Ready-made ornamentation cheapened the product and defeated the overriding purpose of sugar work, which was to demonstrate "craftsmanship."

Setting out to describe how to make not only wedding cakes but also ice sculptures and birthday cakes, Gill expressed ambivalence about the use of mass-produced ornaments, recommending them but then largely circumscribing their use. "A cake may be almost entirely covered with bought ornaments (of which there is an endless variety), such as gum-paste doves, flowers, leaves, etc.; but excepting for the vase and medallions, etc., the less gum-paste used the better when a thoroughly artistic work is desired," he wrote. Ready-made ornaments were best used in cases "where economy is an object, or where the practitioner is not adept in the art of piping."[61] Manufactured gum-paste ornaments were acknowledged substitutes for a confectioner's poorer customer or the unskilled baker, but not for those who wanted "a thoroughly artistic work." "Bought ornaments" signified a lesser cake made by inexperienced hands.

It seems ironic that one so invested in promoting and teaching the skills of fine confectionery would also recommend using prefabricated pieces that diminished artistic skill and turned ornamental cake-making into an assembly-line process. But Gill, like his fellow artisans, maintained a realistic outlook about the general fate of the turn-of-the-century craft trades. Marketplace exigencies allowed little room for the goods of the skilled confectioner. A few worked as chefs and caterers at popular urban hotels, creating one-of-a-kind pieces for the gaze of the elite. But for the most part, mass production and the tastes it generated eclipsed the specialized work of the skilled confectioner.

Even so, the few men who did prosper enjoyed celebrity status by catering to their wealthy patrons. One instruction book even noted people's ability to recognize confectioners' signature creations, much like an artist's paintings or a designer's new gowns, and people of lesser talent trying "blindly" to copy them: "It is a matter for regret that styles are becoming too much 'specialized,' and that the majority of cakes piped by the rising talent in the trade are easily recognizable to the observant as being of so-and-so's style, and are probably either good or bad copies of another man's work."[62] By producing simulations of formerly singular works, confectioners with less skill and imagination co-opted and devalued the original masterpieces. Making copies of celebrity confectioners' work not only heightened the already fierce competition within the trade but also democratized their products, hence softening the class distinctions their existence highlighted and reinforced in the first place. But realistically, most in the confectionery trade knew that monumental sugar works, except wedding cakes (which had gained acceptance as popular ceremonial pieces), did not inspire a steady paying audience.

Even so, many trade journals faithfully continued to include instructions and diagrams in each issue about how to execute ornamental sugar work by hand—both for beginners and the advanced. These articles did not discuss buying ready-made ornaments, but instead painstakingly described how to make various piped and gum paste ornaments like those also being mass-manufactured at the time. The *Bakers Review,* for example, published monthly columns entitled "Ornamental Work for Beginners," which progressed step-by-step through the basic techniques of piping, latticework, calligraphy, and other cake decoration. Gearing it toward general bakers, the editors intended the column to help one expand professionally by providing techniques that would make bakers more versatile and their products therefore more marketable in the trade. The very first column explained this purpose, saying that ornamental sugar work, especially if displayed in one's show window "can create new trade," and men who practiced this art could "make their stores more attractive than can their neighbors."[63] Each month, the articles meticulously laid out plans and instructions that incorporated new designs and techniques, and they always encouraged the beginner to keep trying if he was "not met with success" the first time. Progressing from simple piping work to more complex designs and finally to spun sugar, the directions provided the novice sugar worker with both technical and aesthetic advice. Often columns included suggestions about acceptable color schemes or appropriate design motifs. In this way, the *Bakers Review* taught both skill and artistry, trying to instill into the beginner technical prowess and aesthetic awareness—the two most important possessions of the successful ornamenter.

Other Ceremonial Cakes

By the beginning of the twentieth century, ornamental sugar work commissioned for sculptures fell out of fashion. Yet at the same time, the symbolism of wedding cakes found new expression in other ceremonial cakes. Banquet and holiday cakes, closely related to the wedding cake, accompanied other important occasions and also constituted a large part of the ornamental confectioner's business. Made for Thanksgiving, Washington's Birthday, the Fourth of July, Mother's Day, wedding anniversaries, and birthdays, these cakes marked individually and culturally important events. They also reflected the increasing commercialization of these holidays and functioned as focal points for confectioners' ever-changing window displays.

The birthday cake, for one, only became a viable material manifestation of an individual's birthday when most children lived long enough to be valued sentimentally. The first homemade birthday cake recipes appeared in 1890s' issues of the *Ladies' Home Journal* and the *Boston Cooking School Magazine,* both aimed at the middle-class American housewife. Economic factors coupled with philosophies regarding children determined the relative popularity of birthday cakes. While the working-class families saw their children as productive generators of income, and believed manual labor to be a natural part of their maturation process, middle-class reformers thought of child labor as a disgrace. Dr. Felix Adler, the first chairman of the National Child Labor Committee, insisted in 1904 that "whatever happens in the sacrifice of workers . . . children shall not be touched . . . childhood shall be sacred."[64] Celebrating birthdays with a ceremonial cake reinforced the idea of the "sacred" child among the middle class, making birthday cakes and birthday parties increasingly familiar bourgeois customs, which profited local confectioners.

Birthday cakes and wedding cakes shared a similar material hierarchy; the nature of the cake itself referred to the person it celebrated. Recipe manuals advised professional confectioners to match the cake with the honoree and the purpose of the occasion, advising: "As most birthday cakes are eaten by children, it is well when taking a birthday cake order to enquire as to the ages of those for whom it is intended, and then to select a suitable class of cake. Sponge cakes are often used for this purpose for quite young children. . . . Christening cakes, which are usually for the use of older people, are commonly made of a good rich fruit mixing."[65] Lighter, more delicate cakes suited the palates and sensibilities of younger consumers, while richer, darker fruitcakes were consumed by a more mature (and hearty) audience. Like the bride's cake, which was made out of white cake or angel food in contrast to the darker groom's cake, birthday cakes, too, embodied the symbolic. Celebratory cakes were anthropomorphized and their human counterparts objectified.

Fancy sugar work perhaps most clearly embodied the democratization of refined sugar in America. As ornamental pieces for the dining table or banquet hall, these sculptures embodied male economic prowess in the form of an artwork made of a precious, edible substance that in the end was not eaten at all. People loved sugar first and foremost because it tasted good and had the power to make other things taste good. But ornamental sugar work completely subverted the immediate and practical function of sugar by using it even more luxuriously than in ice creams or bonbons. Borrowing from medieval court traditions, modern subtleties primarily conveyed the idea that the host, an economically powerful man, could demonstrate this both through the use of sugar for such decadent purposes and by commanding someone else's skills to execute such pieces. That the pieces had become feminized to such a degree that the bride and her cake were actually synonymous—a male's possessions in saccharine form—indicated the ultimate cultural acceptance of refined sugar's modern meanings.

CHAPTER SIX

Home Sweet Home

∴ *Domesticated Sugar* ∴

By 1910, Americans were eating on average 83 pounds of sugar a year per capita, a sixteenfold increase since 1811.[1] Confectionery in its various forms accounted for some of this consumption, but many confections—high-class bonbons and sugar sculptures, for example—were not affordable to all people. Sugar's everyday use in the home caused the most appreciable rise in sugar consumption: women incorporated it into pies, puddings, salads, and sauces. In addition, modern food processors made and sold higher numbers of packaged products, many of which came already sweetened.

After the Civil War, consumers began to see sugar as a staple rather than a luxury item. The promotion of sugar and the homogenization of cultural tastes led to increased domestic sugar use. Advertising, home candy-making manuals, and popular women's periodicals linked the home with the marketplace. By 1900, confectionery manuals had transcended their purpose as how-to books, discussing seemingly unrelated topics, such as the role of women, the separation of public and private spheres, and the relative merits of new technologies.

Goods made in the home using refined sugar tended to carry with them the same connections that popular culture had created for mass-marketed confections. "Domesticated," homemade confections conveyed sentiments belonging to the pure and wholesome realm of the feminine domestic environment. Indeed, sugar's ultimate and most natural feminizing processes occurred at home. With the aid of home confectionery manuals and recipes supplied by women's magazines, women themselves had plenty of instruction about the various ways they could incorporate sugar into their families' diets, and the profusion of instructional literature indicates the degree to which Americans had accepted sugar into their lives both physically and psychically. Up until midcentury, Americans relied on a few confectionery manuals for instruction, like the multiple editions published by Raffald, Leslie, Glasse, Price, and Sanderson. But by the 1880s and 1890s, obscure and famous authors alike contributed to the hundreds of dessert, pastry, cake, ice cream, and candy manuals published in America. These later manuals addressed everyone from the genteel woman using her leisure time to experiment with making dainty trifles to poorer women who wanted to make candies out of "necessity" for their families or to sell to others.

In addition to promoting a new fashion, confectionery manuals also reflected the ambivalent relationship turn-of-the-century Americans had with technology and mass production itself. On the one hand, people embraced the new products that technological development made possible; on the other, they remained uneasy about the apparent replacement of people and their labor with wholly machine-made goods. This held true for confections as well. While people devalued homemade candy because it was not a "professional" product made by trained (and male) confectioners or by the sterile machines of large candy companies, they also looked upon it favorably because it represented the personal commitment of an individual's time and labor. Homemade confectionery—a box of fudge, for example—could be both a token of familial affection and a status symbol, indicative of a woman's skill, access to fine ingredients, and amount of leisure time. But it could also represent an admission of economic inferiority and class consciousness, especially in the form of homemade wedding cakes less skillfully ornamented than those bought from a confectioner.

In less than a hundred years, sugar was firmly situated in every American's diet, simultaneously occupying various functions and status positions. In spite of or perhaps because of this, sugar's use and its appearance in popular culture indicated the convergence of many different and often conflicting ideas, not only about sugar and confectionery itself but also about the tensions between public and private, masculine and feminine, wealth and poverty. Unlike confections purchased in stores, homemade candies occupy an advanced place in refined sugar's "biography"—that stage where it was safely ensconced in the home, made safe, domesticated, and largely drained of its more provocative commercial meanings.

Hard Candies

By 1900, sugar had become ubiquitous in American culture and the American diet. Brightly colored, fancifully shaped penny candies could be found in almost every corner store, and children spent countless pennies on them. Reformers worried not only about ruined teeth and food poisoning caused by chemical residues but also about glutted appetites and spoiled characters.

The plethora of candy-making manuals published from the 1880s through the 1910s used the rhetoric of reform to induce mothers to buy the books and to try to make their own candy. These manuals presented as many recipes for good rearing strategies, and definitions of being a good mother, as they did recipes for actual candy-making. Women were induced to buy home confectionery manuals in order to protect their children from the effects of publicly marketed, adulterated goods. The stated purposes of candy-making manuals shifted from promising to produce status items, a chiefly masculine, professional, and commercial enterprise, to urging women to protect their families and fulfill their proper feminine roles of being devoted, responsible family caretakers.

The perfection of sugar production and distribution occurred at the same time that the advertising and publishing worlds saw their most productive output to date. The confluence of sugar refining, food manufacturing, printing processes, and publishing technologies had a profound impact on ordinary consumers. Women's magazines, advertising cookbooks, packaged foods, and popular advice books were all ways in which sugar as a substance and a symbol entered the average American home. For decades, fearing the encroachment of modernization and the results of new waves of immigration, genteel Americans worked to create two separate spheres of life and keep them distinct. Mary Randolph described the goals of such an idealized separation of spheres in the *Virginia Housewife* in 1839, saying, "The husband . . . will feel pride and exultation in the possession of a companion, who gives to his home charms that gratify every wish of his soul, and render the haunts of dissipation hateful to him." Children raised in such an environment ended up being moral and "of steady habits," and daughters particularly, "if the mother shall have performed the duties of a parent in the superintendence of their education, as faithfully as she has done those of a wife," would mature to raise proper families of their own.[2] Advisors characterized the public sphere as a world filled with strange, foreign people and temptations, "haunts of dissipation," driven by cut-throat capitalist pursuits. In contrast, the inner domestic sphere was a protective and protected sanctum, which nurtured the spiritual growth of the family. Men ventured out into the public sphere to earn a living, while women kept the domestic sphere sealed off from the negative influences of the outside, which enabled them to raise their children to be good citizens.

Yet nineteenth-century Americans found it impossible to keep these spheres distinct. In fact, many women adopted new goods and services delivered by technology *because* they believed it would help reinforce the household domain and protect their children from the evils of the marketplace. Paradoxically, the very media that promised to keep the public sphere at bay were also the ones that introduced marketplace goods into the home. The homeward-bound flow of sugar, one clear example of this process, shows how the home, rather than being an impenetrable barrier to the marketplace, actually facilitated commodity exchange. By the end of the nineteenth century, home candy-making manuals touting particular lines of candy-making equipment or promoting local candy-making schools had become wildly popular.[3] In 1883, one author pointed out that in the recent past, "home-made candy was still limited to a few varieties of molasses candy, [with] only the ambitious girl venturing upon caramels or drops of any sort. To-day it is not only possible for any one to make excellent candy for home consumption, but even to imitate successfully the choicer varieties of French candy."[4] But as cultural products, these publications reflected the irreconcilable impulses that drove people's desire to keep the public and private spheres separate. The marketplace offered an ever-increasing supply of novel goods delivered by mass production. Yet at the same time, people looked upon the marketplace as a dangerous, unpredictable realm, where strangers interacted with one another and with new, sometimes alien consumer goods. "A genuine 'privatization' of family life never occurred," T. J. Jackson Lears says of this unavoidable paradox of late-century American life. "It was impossible for the home to remain altogether isolated from the market society. Inevitably, the haven embodied many values of the heartless world outside. If the home was meant to be a refuge from the marketplace, it was also meant to socialize people (particularly males) to succeed in that competitive realm."[5] In fact, the home became the main context in which the family consumed marketplace goods and integrated them into their daily lives. Home confectionery manuals both expressed and tried to reconcile these contradictions. Their sentiments merely reflected the public/private paradox that troubled turn-of-the-century women in particular: women could only protect their children from the vagaries of the public sphere by incorporating its products into their homes. The rhetoric of home candy-making instruction books expressed this paradox repeatedly.

Because mothers loved their children and children loved candy, authors of candy manuals explicitly addressed these relationships in their writings. Many authors claimed that their sole purpose was to provide simple home recipes enabling mothers to give their children unadulterated versions of the penny candies they so often craved. Rather than wean children off candy altogether, women might do better to create homemade batches of pure candies. "There is no doubt that Americans eat too much sweet-stuff of one sort and another," wrote one author, "but as it is a national weakness, it is a good thing to know the purest forms."[6] Manuals touted the benefits of their recipes in the familiar terms of other prescriptive literature of the time. They often included prefaces and introductory remarks that reminded women of the dangers of the outside world and their obligation to protect their children from those dangers—a diffuse fear of the marketplace most succinctly expressed in terms of the potential hazards posed by adulterated candies. Popular critics' familiar concerns about penny candies were advanced in these manuals as reasons for women to make their own sweets at home. Since children could not be stopped from eating sugar, the manuals argued, they might as well eat candies that mothers could be sure would not harm them.

While some popular commentators believed that sugar was not only healthy but also nutritious,[7] others outwardly disapproved of children's candy consumption. In contrast to marketers who promoted confections for their own business interests or on behalf of the sugar industries, culinary experts like the respected Sarah Tyson Rorer called sugar "evil," and counted its consumption as one among a number of "dietetic sins." In a 1906 issue of the *Ladies' Home Journal,* she claimed that cravings for sugar resulted from "the ravings" of a "deranged" and "insane" stomach. She thought women's desires for sweets posed particular dangers, and blamed them for "the various diseases to which they are subject." Like other reformers of the time, Rorer also held mothers responsible for their children's sweets cravings, saying, "Our dietetic sins are started in our very youth." She presented an "average" scenario for the reader, involving the mother who uses "a liberal amount of sugar" on her child's breakfast cereal. "Such children know nothing else from their very earliest recollections," she explained. "When the mother is told to eliminate sugar the answer is always the same: that the child will not eat anything but sweets, with this explanatory note, that 'he has always had such a craving for sweets.' Why not? He has been taught from birth to eat sugar."[8] For people like Rorer and her followers, excessive sugar intake constituted bad mothering rather than simply dietary miscalculation. Maternal indulgence (reflecting the mother's own craving for sweets) endangered children. Such mothers started their children on the road to perdition at an early age.

Debates about sugar consumption in the home defined and circumscribed the proper roles of women. Christine Herrick, a popular health and economics advisor, echoed Rorer's sentiments about the "evil wrought by this ill-directed love of sweet stuff," saying: "Children are allowed to buy and eat candy as they please," by "parents who would shrink in dismay from permitting a child to touch a labeled poison. . . . The small child is given his penny or nickel, when he begs for candy, and is turned loose to buy where he will."[9] A loving, caring, good mother properly oversaw her children's diets and did not indulge them. Their intake of sweets became not only an issue of good health but one of morality as well: the degree to which a woman gained control over her children's desires—for sugar or anything else—indicated how successful she had been in civilizing them and making them fit for the public, adult world. In the end, keeping children properly confined to the domestic sphere constituted thorough parental training. Politically, it ensured a stable society grounded in and based on the ascription of stereotypical gender roles, making the mother responsible for raising her children to be good and productive citizens who would also assume those same gender roles themselves as adults. By 1900, the typical confectionery manual, born from the unlikely marriage of parenting manual and recipe book, once again linked women, children, and sweets in a seemingly natural way.

Dietitians also entered the fray, championing candy as an "energy food," a position that only slightly assuaged the concerns of the pure food advocates, who saw any food on the open market as a threat to the moral and physical fiber of the American people, and specifically to children. Regulating sugar consumption was secondary to the protection of future generations. Dietitians concerned themselves with safeguarding the general health and well-being of America itself against the enervating effects of candy and the dangerous effects of adulterated candy in particular. A woman who really cared about the welfare of her children, the manuals insisted, should make candies herself rather than barring them altogether. One turn-of-the-century cookbook expressed this clearly: "If candy is taken under such conditions that it will not derange the digestive apparatus, it is perfectly wise and rational to be a candy-eater. And if candies are to be eaten, those prepared at home are sure to be free from injurious ingredients."[10] According to this author, "rational" candy consumption—that which did not happen on impulse and, ostensibly, was devoid of pleasure and spontaneity—qualified as an acceptable, unthreatening activity.

Middle-class women enjoyed making candy themselves and included it among their genteel leisure activities. They satiated their children's cravings for sweets while guaranteeing that the sweets were pure, obviating the need to send the little ones outside, where "they strike for the nearest shop or store, and where they can get the most for their pennies."[11] Another author wrote that "candymaking is a pleasant pastime that serves to occupy the attention of the young at times when other pursuits, far more dangerous, might be indulged in."[12] Of course, rich people could always afford to buy what they thought to be the best and purest candies in high-end establishments: "It is not supposed that there is no pure confectionery. Those who purchase at our best and old-established places, are morally sure of getting a genuine article."[13] However, children relying on their nickel-and-dime budgets turned to less expensive varieties, necessarily made from cheaper ingredients. Making candies in her home, an affluent mother supervised the welfare of her children by averting unpredictable and unhealthy purchases, which might invite more dangerous addictions to tobacco or alcohol.

The influential *Boston Cooking School Magazine,* which began publication in 1896, took a more rational approach, advocating a balanced diet for the family: "Sugar is needed to supply heat and energy, but the dietaries in most families and institutions are apt to contain too much starch and sugar and too little protein."[14] Eating a variety of basic foods in moderation maintained a sound body and mind; conversely, certain foods ingested in excess disrupted one's physical and psychological stability. For example, the magazine forcefully warned women, specifically, about the dangers of eating too many sweets during the holidays, not because it would endanger their health, but because it would affect their emotional state and cause symptoms indicative of addiction and withdrawal: "If we are to have the health that brings brightness to the eye, elasticity of step, and cheerfulness in manner, we cannot eat anything or everything that an unrestrained palate may fancy. When nibbling the rich pastry and cloying sweets of Christmastide it is well to keep this in mind, if we mean that the aftermath of Christmas shall not find us irritable, gloomy, and taciturn."[15] Because of sugar's status as a culinary mainstay in middle-class holiday celebrations, the magazine's advisors could not realistically prohibit the intake of all sweets. But they could and did warn of the consequences awaiting those with "an unrestrained palate"—the "irrational" consumers prone to overindulgence. Almost always women and children, they could not be trusted to control their own appetites. Sugar had become such a prevalent ingredient in the American diet that the caveats about eating it seemed to equal the number of recipes for using it. Yet in the face of all the criticism and all the warnings, Americans continued to eat sugar in ever-increasing amounts. In 1801, Americans consumed 21,376 tons (42,752,000 pounds) of sugar; in 1905, they consumed about 2,632,216 tons (5,264,432,000 pounds) of refined sugar alone, a figure that does not include other sweeteners like molasses and honey.[16] This meant an increase from approximately 8.4 to a staggering 70.6 pounds of sugar per person per year, certainly a long way from a trickled spoonful of the stuff in a cup of tea.[17]

Mary Elizabeth Hall's *Candy-Making Revolutionized* (1912) attempted to answer concerns about nutrition, purity, and cost-effectiveness all at once by teaching people how to make confectionery out of vegetables alone. Instructions included making fondants and pastes from potatoes, sweet potatoes, parsnips, carrots, beets, tomatoes, and green beans. The book even included a recipe for lima bean taffy. Hall was probably closer to the truth than she imagined when she claimed that her vegetable candy "furnishes the valuable element of sugar so combined with nutritious vegetable bases that, because of the bulk, *there is no temptation to overeat.* This quality of new confection would seem insurance against the evil effects of gluttony! Before an undue amount of sugar is consumed, the very mass of the vegetable base has satisfied the appetite."[18]

Books explaining how to make candy from vegetables notwithstanding, the popular literature did serve to homogenize the American palate not only by bringing mass production into the home but also by bringing urban culture to outlying areas, including places like Auburn, New York; Canton, Ohio; Elkhart, Indiana; and Hartford, Connecticut. G. V. Frye attested to the confectionery deficit in outlying areas, writing that "it is often difficult to procure goods that are fresh and pure in our smaller cities and towns; hence they must either send for them, which is expensive, to say the least, if not mostly unsatisfactory, or buy the stale and often impure goods handled by the grocer or small confectioner."[19] Not only did homemade candies (symbolically, at least) mark the distinction between public and private spheres, but those produced in smaller towns bridged the gap between the abundance of goods in urban places and their dearth in rural areas. In both cases, people constructed ideas about self and other, safety and danger, familiar and unfamiliar through the availability of goods in their towns and in their homes. Making candies oneself limited the encroachment of the outside world, while making its amenities accessible at the same time.

Class distinctions paralleled geographical divisions. In elite homes, where upper-class children lived longer and enjoyed more hours of leisure, fudge-making and taffy-pulling parties were all the rage at the turn of the century, and a birthday party was incomplete without ceremonial cake and ice cream. While fathers worked all day in the public sphere, upper- and middle-class mothers oversaw the needs of their children, assisted by retinues of domestics. In addition to providing for their basic needs, mothers furnished their children with amusements that would stave off boredom, satisfy their desires, and perhaps teach them principles about the public world that they could use in the future as grown-ups.

Because children by the turn of the century had developed their collective sweet tooth to such a degree that the desire for sugar was ingrained, using confectionery and candy as teaching tools disguised as sources of entertainment perfectly channeled physical appetites into instructive experiences, further "rationalizing" desires. *Our Reliable Candy Teacher* was confident that "we are filling a long felt want in issuing this book." Recognizing "that it is the parent's duty to have something in their home to entertain their children or they will go out in the world for pleasure," the authors recommended using the book to teach about commerce, because "although they may never go in the confectionery business, it will be a lesson worth learning."[20] Things relating to sugar piqued children's interests, and candy-making manuals promised that home confectionery would focus their attention enough so that they would not look beyond the home for pursuits of pleasure. Even better, candy-making activities encouraged children to be together. "How many of my readers have tried the making of fudges as an innovation to the romping and 'do-nothing' method in which many of our young folks spend their time when companions come to while away an hour?" asked one author. Taffy-pulling and fudge-making parties became events that fostered like-minded sociability among genteel children, preparing them for future success as cooperative yet ambitious adults in the public world: "'Candy pulls' are among the most enjoyable events of children's lives. . . . Imagine, if you can, some quarter of a hundred youngsters repairing to the kitchen all bent on doing their share to make the occasion what is implied by the words, 'Fall in and help yourself.'"[21]

For women, the practical aspect of making candy in the home ensured their children ate purer forms of sugar. But the process also conferred status upon the house. For the refined, candy-making exemplified Thorstein Veblen's "vicarious consumption" and "conspicuous leisure." A woman not only flaunted her free time through such pursuits but also functioned as a man's surrogate consumer. "The housewife's efforts are under the guidance of traditions that have been shaped by the law of conspicuously wasteful expenditure of time and substance." Moreover, "the wife, who was at the outset the drudge and chattel of the man, both in fact and theory—the producer of goods for him to consume—has become the ceremonial consumer of the goods which he produces."[22]

Demonstrating that you had the time to trifle in the kitchen making such nonessential things proved much more important than creating perfect candies. As one author put it, "To those who have leisure and the means to indulge their innate love of cookery, sweet-meat-making can be carried to a fine art, for wonderous [*sic*] are the productions made of spun sugar and of boiled sugar in varied forms."[23] Another author noted, "The modern elegant devices by which strawberries, violets, and orange-blossoms are candied in sugar, effect a Home Amusement for dainty-fingered girls."[24] This kind of candy-making, indeed an indulgence, required time, supplies (sugar, flavoring essences, coloring extracts, and equipment), and the suitable social opportunities at which to show the results off to a like-minded group of peers. Because sugar was sweet and delicate, people equated it as a material with the inner nature of the refined woman. Making delicate candies qualified as a suitable activity for delicate hands, not a "ruder" job like cutting meat, meant for stronger hands, as one 1866 manual pointed out perhaps earlier than any other. Many women judged other women on the quality of their homemade sweets and desserts, thereby making homemade confections not merely tasteful dainties but, more important, crucial morsels that defined and determined social standing and degree of refinement: how you were with sweets indicated how sweet you were yourself. Even a Jell-O advertisement of 1907 promised that with Jell-O powder, women could make "something that more or less critical feminine company will discuss favorably."[25]

Domesticated candy-making had a practical side as well. For women needing to make their own livelihoods, domestic manufacture of sweets often resulted in a viable business, and could supplement other forms of income. Some women actually did go into business themselves, running small confectionery shops out of their own homes. An 1860s book, *How Women Can Make Money,* detailed women's opportunities in "fancy confectionery," but only listed the piecework they did (wrapping candies, glazing bonbons, packing boxes), saying, "Making common candy is said to be too hard for women."[26] But women actually did run a number of the local businesses—some operated branches of their husbands' stores, and others worked as independent confectioners. According to Philadelphia directories, women ran 13.8 percent of the city's confectionery shops in 1850; 20.23 percent in 1859; 26.1 percent in 1874; and 33.5 percent in 1881.[27] While upper-class women made confectionery to amuse themselves and impress their friends, some middling and lower-class women found it a profitable endeavor, sometimes setting up shops right in their own homes or selling to local drugstores and candy shops.[28]

If one made the "choicer varieties of French candy," for which "there is always a certain sale," candy-making was both amusing and profitable, *The American Girl's Home Book of Work and Play* (1888) suggested. "Its preparation requires time, patience, delicate handling, and the skill which comes from even a short practice in the use of these prime essentials. . . . a very comfortable sum could be made monthly by supplying the drug-store or the village store with the carefully prepared and pretty bonbons."[29] Of course, only those homemade candies that looked like "choicer varieties," resembling professional goods already on the market, and those only for small-town drugstores or grocers, qualified for sale to the public. Another manual promised that with practice, one could make "Orientals," "The Finest Chocolate Cream Made," and that because these were hard to find and perishable, one could turn a sizable profit by filling custom orders, promising, "you will have no trouble in selling all you can make at sixty cents per pound, to private customers only, as there are very few stores in the country where it is possible to purchase them. One reason of this is, they are too delicate to stand being boxed up and shipped around the country to the different dealers, and probably be kept for months, as some candies are, before being sold."[30] Very few women actually did rely on candy-making businesses for their sole support. Breaking into this historically male-dominated profession remained difficult, because of the reputations of established French, German, and English confectioners. Even city directories recording a significant number of female confectioners reveal a sizable turnover rate from year to year. Women made and sold candy primarily to supplement other income, or remained dipping and packing girls in larger candy factories; those who tried to rival the fine confectioners could not escape the stereotype of being mere amateurs or hobbyists.[31]

Candy-making manuals, like other prescriptive literature of the time, tried to elevate, to "re-form," the tastes of the working classes, but at the same time they also recognized the limited effectiveness of such endeavors. In *Home Candy Making* (1889), Sarah Tyson Rorer did not promise to teach making luxury items, but instead aimed at what she believed could be accomplished by the less able and less fortunate. "This little book is the result of careful practice in teaching beginners how to make attractive, wholesome, and palatable varieties of home-made candies . . . and, while they may be palatable to some persons, to the connoisseur they are coarse and heavy," she wrote.[32] Rorer made no attempt to elevate the tastes of "the masses," as she called her audience, but instead sought to satisfy their wants by providing them with simple, basic confectionery recipes. Her manual, like most others, encouraged aspirations to higher classes at the same time as it reinforced class hierarchies: instead of suggesting that people do without such frivolities as saccharine treats, popular authors provided ways in which the poor could have access to them, thereby advancing the idea that consumer desires, however misguided, should be sated rather than ignored.

Ice Cream

Another sugar-centered leisure pursuit that occurred in the nineteenth-century domestic sphere with increasing frequency, ice cream making often involved the entire family. It was the one confectionery-related activity in which men actively participated, mostly because it involved hard physical labor and the use of a heavy piece of machinery—the ice cream freezer. The popularization of the ice cream freezer sparked the popularity of home ice cream making. Before the advent of the freezer, patented in 1840, making ice cream required a large metal pot filled with ice and salt, which held a smaller pot filled with "tea" (cream, eggs, sugar, flavorings). The inner pot had to be continually rotated by hand even to achieve freezing, and frequent stirring was needed to render its contents creamy. The ice cream freezer combined the inside and outside pots into one apparatus, which was outfitted with inner paddles and an outer crank. Rather than juggling overflowing bowls of ice and sloshing ingredients, one could, with some effort, neatly accomplish the job of freezing ice cream to a smooth consistency with this new machine. The liquid ingredients were contained in the inner chamber, which sealed at the top to prevent their contamination by the surrounding ice and salt mixture. The insulated outer chamber conveniently held freshly shaved ice and accommodated the addition of more salt to expedite the freezing process. Turning the hand crank rotated the paddles, producing a smooth, fresh, finely textured product reminiscent of that sold in ice cream saloons.

Some people have mistakenly claimed that ice cream freezers and ice boxes were common appliances in all homes both before and after the Civil War.[33] Only the rich purchased the novel appliances and had the time, inclination, and raw materials to make their own ice cream. By the end of the nineteenth century, ice box and refrigerator ownership varied greatly by class and region. "It is difficult to estimate how many people had refrigerators," Oscar Anderson observes. "One writer asserted in 1884 that they were as common as stoves or sewing machines in all but the poorest tenements, but there is no doubt that boxes or tubs often served as substitutes, and that poorer families used ice only in small quantities in the hottest weather if at all."[34] By century's end, ice itself was certainly more accessible, but much of this ice went to supplying commercial butchers, hoteliers, hospitals, and professional confectioners, making home use in urban areas an especially unusual occurrence.

But by 1880, home ice cream freezers for the middle classes had become popular enough that the U.S. Stamping Company, for example, not only offered "White Mountain" freezers but also published a 25-item price list of replacement parts for the freezers, which included everything from a new gear frame to a replacement beater. The fact that there was a market for both new freezers and spare parts suggests a busy trade in such machines. "A good ice-cream freezer is a valuable accessory in any household," George Peltz, the editor of *The Housewife's Library* (1885), said.[35] Storing ice cream until ready to serve required an ice box.[36] For such purposes, Peltz recommended the "Empress," which was an "ice-using refrigerator" fit for "domestic uses" that looked like a finely crafted gothic revival sideboard. The "Snow-flake," proportioned like a baby's high chair, was "another ornamental form, intended for use in the nursery."[37]

The increasing prevalence of the ice cream freezer in middle-class homes necessarily stimulated the proliferation of recipes for individual use. Many instructional freezer pamphlets included ice cream recipes in order to make consumers feel confident that they could successfully use these devices. Instructions for making ice cream also appeared in women's magazines, domestic cookery books, and separate treatises on the subject, and, like candy-making, it had become a fashionable hobby by 1900. One author even wrote a "Sermon on Ice Cream," which decried the commercial production and consumption of inferior ice creams, and in particular criticized street vendors and local grocers, who were "conscienceless chemists" providing "disgusting substitutes" for "God's own provisions."[38] Many authors agreed that guaranteeing pure ice cream, unadulterated by inferior ingredients and untainted by verdigris,[39] meant making it within the sanctity of one's home, echoing the popular sentiments of the home candy-making manuals' authors.

Domestic ice cream production and consumption counted as near-sacred activities for the family. Confectioners were reluctant to deliver on the Sabbath, so families who wanted ice cream after their Sunday dinners had to look elsewhere. Making ice cream on Sundays and holidays like the Fourth of July thus became a common practice for the middle classes. Cornelius Weygandt, writing of his childhood in the 1880s, described the "ritual" of making ice cream on Sundays, which started with amassing all the requisite ingredients and supplies and culminated in the "ceremony of 'licking the paddle.'"[40] Then, after each family member had sampled the ice cream by the spoonful, they adjusted the sweetness and flavoring and repacked it with ice and salt to "ripen" before serving at dinner.

Even though home production helped democratize ice cream, class distinctions associated with it remained: even versions of homemade ice cream differed in appearance and quality depending on one's particular class. The middling classes most likely "put up" their ice creams in ubiquitous steel melon molds. In contrast, the rich served their homemade ice creams in fanciful shapes formed using elaborate molds of tin-plated copper or cast iron. Often imported from Britain, the molds enabled the rich to emulate the professional confectioner's product. Machines stamped out steel melon molds in one action, sacrificing height and stylistic variation for mass production. Imported copper molds, however, which were still handmade, assumed more elaborate shapes, and the ice creams they formed were important status symbols when entertaining guests, literally leaving the impression of the costly mold, in ice cream, for all to see.[41] Cornelius Weygandt recalled the centrality of fancy ice creams in marking certain holidays, demonstrating how the elite established this as a fashionable custom. Since his family lacked the shaped molds to turn out fancy ice creams properly, they sometimes went to the local confectioner. "For parties, we got the ice-cream at Harkinson's," he wrote. "And a party was not a party if there was not ice-cream. . . . Ice cream was the *sine qua non* of Fourth of July and Thanksgiving, to have a whole little turkey of ice-cream set upon your plate; or, at Christmas, a little Santa Claus; or, on the Fourth of July, a flag or a cannon!"[42] Ice cream, and especially shaped ice cream, properly concluded such celebrations, as essential as the period at the end of a sentence.

Upper-class homes had an assortment of molds, some for ice sculptures, some for ice cream, and others for fancy cakes. Domestic advisors considered them necessary equipment for well-appointed kitchens. Many wealthy families, in fact, commissioned customized molds that imprinted anything from family crests to pets in the frozen ice cream, creating chic desserts influenced by those served in the public sphere. It was one thing to attend a catered banquet with food provided by a professional; it was quite another thing to provide such fancy food oneself, and the refined woman continued to do so. The *Boston Cooking School Magazine,* for example, featured many articles for the genteel woman on making ices and ice creams. In its 1896 premiere issue, the magazine published a five-page article on "Ices," explaining their history and providing recipes. "No form of dessert is held in such high esteem as the frozen; it is, at once, the best approved and the most palatable of all the desserts," the author declared. "By the presentation of a well prepared cream of good quality, one may atone for a very plain dinner."[43] Appearance and content vied for importance in these upper-class desserts. Their shape, and the degree to which their ingredients were manipulated ("culturized"), determined their degree of luxury. These desserts embodied a housewife's domestic abilities in the kitchen, serving as tangible manifestations of her husband's economic prowess (via fancy molds) and demonstrating possession of both the time and the ingredients necessary to turn out such fine confections.

Bernard Lyman observes that the shapes of foods greatly influence our perception of them. Uneven spoonfuls of ice cream, for example, "convey a homemade feeling," whereas "complex, unnatural forms are more likely to be pleasing for special effects, especially when one wants to make a point that great care and effort went into preparing the food."[44] Ice cream for parties had to be in molded shapes to capture the spirit of the occasion, but ice cream made and served within the family could be dolloped right from the canister. One could even gauge the "class" of the article based on how controlled and managed its form was. Ice cream in blobs connoted casual eating situations, congeniality, and familial bonding, while ice cream in ornamental forms connoted formal, ritualized occasions.

Ice cream in molds recalled the skill of the confectioner, a man of class highly respected among his neighbors and regular customers. A woman who served her own ice cream attempted to imitate the confectioner and to assume some of his cultural currency and status. One woman, writing to the *Boston Cooking School Magazine* in 1897, asked how she could get her molded ice cream to come out with "the smooth consistency seen in that which I have eaten away from home."[45] This prompted a page-long response from the columnist, detailing the various complicated processes used to make mousses, parfaits, and glacés like the professionals. If women wanted to impress people with their culinary skills, they needed to take their instructions from the most revered sources.

Manuals of all sorts simultaneously helped to democratize both the processes used and the desire to make ice cream by patiently explaining the various ingredients required and giving sometimes as many as a hundred different recipes for ice cream variations. Domestic scientists authored many of the booklets that, as promotional and instructional materials, accompanied new appliances. For example, Rorer compiled *Frozen Dainties,* which contained recipes emphasizing the ease of making ice cream with the "White Mountain Freezer." She also contributed recipes to the booklet *Freezers and Freezing,* for the "Lightning Freezer." The various experts—spokespeople—assured people that making ice cream in the home would "prove much more satisfactory than the 'brick cream' which is often made in an unclean basement, by careless and unwashed employees."[46] Here again appeared the contradiction between isolating the domestic sphere from possible outside contaminants and a concomitant marshaling of technological components from this outside world to do so.

In addition to freezing apparatuses, new food products also brought ice cream into more and more homes. Having developed crystallized gelatin and ice cream powders at the turn of the century, many companies offered cheap, convenient products that promised to produce fine imitations of fancy ice cream, allowing the middle and working classes to imitate upper-class creations. Companies took the same principles used in their granulated gelatins and applied them to making ice cream "powders." Selling at 10 cents a box, these powders, when mixed with milk or cream and then frozen, yielded a batch of "ice cream" (which must have had a texture like frozen pudding). These powders traded on familiar sentiment that assured the consumer of goodness through technology. A 1902 advertisement for White Frost Ice Cream Gelatine stressed its "purity" and "delicacy." D-Thicka, an ice cream additive, claimed to be free from "any article prohibited under the Pure Food Laws." And many other new products suggested purity and wholesomeness through their names alone, emphasizing lightness, whiteness, and powdery qualities that moved from the concrete to the ephemeral: Crystal Flake and Velvouette Ice Cream Powder, for example.

A few years after its inception, Jell-O also began making its "Ice Cream

Powder," in Vanilla, Strawberry, Lemon, Chocolate, and Unflavored. Costing only "about one cent a dish," this "preparation" promised an affordable treat for the ordinary working classes. Jell-O's advertising copy scoffed at the traditional way of making ice cream, and took what for the elite constituted a quaint leisure activity on a summer Sunday afternoon and turned it into an outmoded, inconvenient, antiquated task: "The old way of making ice cream, with its toilsome and expensive process of beating eggs, mixing, sweetening, cooking and flavoring, was slow, uncertain and unsatisfactory. After fussing for an hour you never knew whether your ice cream would be eatable or not, and usually it was not." Jell-O celebrated its technological origins, and modernity in general, by implying that women remained slavishly loyal to inconvenient, unreliable ways of making ice cream. Indeed, the very appearance of Jell-O's advertising booklet illustrations—brightly colored, depicting the gelatinous dishes in lush, luxurious settings and placed on silver platters or gold-rimmed dishes—turned Jell-O Ice Cream Powder into a product delivering the good life.

Ice cream powders offered convenience and efficiency—traits especially important for those with little time and equipment. Technological wonders, instant powders created fancy desserts simply, in just the few hours it took to freeze them. The poor could now enjoy middle-class affectations: "As everyone likes ice cream, and every hostess considers it a necessary part of the menu *for a formal function,* it is important that the simplest and most economical method be adopted for obtaining it."[47] Technology aided by advertising had transformed ice cream from a special family activity into a "necessary part" of important meals. The directions for Jell-O Ice Cream Powder were similar to those of other brands: "Add a quart of milk (or milk and cream mixed) to the contents of one package and freeze. NO HEATING OR FUSSING."

Instant powders made accessible formerly exclusive, elite goods and blurred the class distinctions that had been clearly and materially expressed. Now, even poorer women competed in the game of conspicuous consumption by impressing their friends and families with good-tasting and good-looking desserts. Advances in food technology displaced the ritual aspects of ice cream making, which had always been a time-consuming and family-oriented process. Of course, poorer women did not have fancy molds to completely pull off such upper-class creations, but food companies offered alternatives via advertisements. For example, Jell-O Ice Cream Powder suggested using a leftover baking powder can to mold its "Philadelphia Ice Cream." Even better, they instituted a premium system allowing one to exchange a certain number of product wrappers for a melon mold. Jell-O, among other companies, recognized the importance of food aesthetics, stating in one of its advertisements, "Fancy Jelly moulds are prized by all housekeepers who take pride in the appearance of their dining tables, but they are expensive and hard to get. As many people have written us that they were unable to buy them in their towns, we have had a large quantity manufactured. They are made of aluminum in beautiful patterns."[48] So the Jell-O company and other similar manufacturers of the time not only influenced American foodways themselves, but also, through the manufacture and promotion of accessories like molds, homogenized the appearance of food, taking away much of its symbolism and replacing it with commercially created and disseminated mass-produced standards.

At the same time that high technology geared up to democratize luxury items by producing them faster and more cheaply, high culture shifted to reemphasizing the things that made it elite and distinctive to begin with. The wealthy, shunning the new convenience products, favored fancy dessert molds, and tedious, time-consuming recipes. Now that the disenfranchised attained a piece of high culture by making "fancy" desserts, the elite's desserts became even fancier. Although access to the time and equipment to mold ice cream tacitly had class connotations, aesthetic considerations reigned supreme. The elite had access to greater numbers of fancier, more ornamental, decoratively encrusted objects and commodities, while the poor settled for simpler versions. As important, material differences in the foods themselves distinguished classes of goods. Simply put, homemade ice cream from the best ingredients must have been richer and more satisfying in taste and texture than that mixed from a box of chemically formulated powder.

Chocolates and Soft Candies

While homemade ice cream simultaneously symbolized high living and familial bonding, homemade chocolates remained firmly rooted in the creation and maintenance of social networks. Costly, commercially made chocolates indicated romantic love and carnal desire. Typically not made by amateurs, they required skill that came only through repeated practice, and the costly, perishable ingredients had to be treated with care. In contrast, homemade chocolate fudge, which could be made quickly and cheaply, came to represent a domesticated love—a sentimental gift shared between family members or among close friends. "The moods or feelings that candy produces are rooted in memories and associations: the home, parents, school days, happy occasions, a reward for some good deed," one author observed. "Love of home and of parent and child, with all comforting connotations, is therefore the deepest motivating force about candy."[49] Gifts of homemade fudge, for example, moved among familial or social intimates and gained their greatest meaning because they resulted from a woman's own labor. "Professional" looking treats, in fact, may have had less of an impact on the recipient of the gift than more naïvely fashioned confections, which were "especially for the delectation of the home circle, where the quality and pleasant flavor of the product is sought, rather than its manufacture into any particular form."[50]

Gifts, singled out from the universe of generic, mass-produced goods, served as important "tokens" of friendship, affection, and remembrance. People attached a great deal of sentiment and meaning to humble gifts, particularly, because of their contrast to mass-produced goods in the marketplace. "It often takes courage to give little things,—a cake of maple sugar, a butterfly bow, a single rose of cheap variety,—lest the gift seem so insignificant that it ought not to have been given," one periodical writer commented. "Nevertheless, it is just such small offerings that make one feel she is remembered; and remembrance cements friendship."[51] Chocolate in the form of homemade fudge, divested of its associations with connubial love, instead conveyed a sense of warmth and friendship. In this form, it represented time spent in the kitchen rather than economic ability.

People who did not give fudge and other similar homemade candies as gifts often sold them, but only in domesticated marketplaces. "Fudge tables are very attractive and yield handsome profits at church fairs and bazars [*sic*] and they do not require elaborate preparations," noted one writer.[52] Another author confirmed the success of homemade candies in feminized sites of commerce: "At a bazaar, confections that are home-made always find a ready sale."[53] For this purpose it was not necessary to make confections that resembled fancy bonbons or had upper-class aspirations. In fact, consumers preferred unprofessional-looking goods that better suited domesticated markets that valued handmade individuality and old-fashionedness over mechanized uniformity.

By the time most handmade goods had been superseded by machine-made commodities, even some large candy companies tried to capitalize on the nostalgic appeal of home and hearth. In 1912, Whitman's Chocolate Company introduced its signature Whitman's "Sampler" line, a "personalized selection"[54] of semi-fancy, soft-centered candies packed in a box decorated with faux needlework lettering. This was an ingenious piece of marketing—a company producing thousands of boxes of chocolates a year capitalized on the home candy-making craze by marketing its product in a way that tapped into the nostalgia for handwork and better days.[55] The needlepoint sampler served as an apt motif for Whitman's to appropriate because people immediately recognized it as an icon representing the quaint and innocent domestic harmony of the romanticized, mythic past. As Whitman's advertising has claimed, "Americans have . . . *loved* the Sampler for what it says—in sentiment and in compliment."[56]

Kenneth Ames has written about the popularity of needlework samplers in American culture during the 1870s and 1880s, observing that in form and message, they were powerful ideological communicators of "charged and resonant concepts, values, and beliefs."[57] More specifically, Ames locates these meanings in a comfortable past firmly grounded in religious tenets: "Most of the mottoes seem either linked to or at least consistent with evangelical Protestantism of the period. They expressed the values of the conservative center."[58] It is no wonder, then, that Whitman's chose this decorative theme for its box, "done in the old-fashioned style, with authentic sampler motifs,"[59] for it called to mind many generations of homespun sentiments, suggesting, if not overtly referencing, the sacred. The Sampler was a down-home, domesticated version of rich bonbons and decadent chocolate-covered cherries—a much easier product to swallow metaphorically.

Many domestic instruction books on candy-making traded on the sentimentality of the homemade by including directions for making one's own packaging, noting that presentation equaled in importance the goods themselves. While the rich might pay $50 or more for a fancy French bonbon box, the poor and middling created their own imitations of fancy containers using the residues of mass production. For example, one candy manual described ways to package homemade candies tastefully by recycling manufactured goods: "Candy boxes may be bought in almost every town, but if you have saved some that you have received, these may be used as well. Paste an appropriate postal card over the name of the firm on the lid."[60] Another suggested: "A great variety of bon-bon boxes and bags may be home-made, especially by any lady who is neat-handed with her needle and who can paint. . . . Little cardboard boxes such as one often gets when purchasing note-paper, jewellery, &c., can be ornamented with painting small playing-cards, artificial flowers or butterflies . . . and any other fanciful designs."[61] Techniques like these enabled women to imitate upper-class goods and to further reinforce the personal homemadeness of their goods. By taking leftover postcards, clippings, and boxes—castoffs from commodity culture—and incorporating them into gifts, one could create something unique, original, and individualized.

Nostalgically domesticated confectionery also manifested itself in the sudden popularity of "old-fashioned" candies like "Farmhouse Fudge,"[62] recipes for which appeared often in candy-making literature. Pieces of fudge, molasses chews, and peppermint sticks, these candies hearkened back to the "old days," appealing both to mothers looking for safe candies and also to the elderly, who may have remembered treats with such medicinally inspired ingredients from their younger days, when the mass popularity of candies was just beginning and they were sold by druggists.

As mass production overtook the confectionery industry, forms of sugars and flavorings with less cultural cachet better succeeded in conveying the messages of home, hearth, and the warming glow of tradition. Molasses, for example, a very early sweetener in America and common by-product of refined sugar, was used frequently in the South by the poor (and slaves), who could not afford white, crystallized refined sugar. The December 16, 1834, issue of the Philadelphia *American Sentinel,* for example, featured a recipe for "Mrs. Wigmore's Molasses Candy" that called only for brown sugar, lemon juice, and a quart of West Indian molasses. After the Civil War, when technological innovations and the influx of immigrants changed the cultural landscape of America, Anglo-Americans commonly and fondly looked back on their real and mythic pasts. Since confectionery, like other goods, reflected current and changing fashions, nostalgic impulses influenced its nature, too. In 1875, a writer praised molasses candy as "a title of merit," used by the confectioner "to indicate the similarity of his products with that made by our grandmothers in the days when French *bonbons* were a rarity seldom seen outside of the very large cities." Chronological distance replaced rarity, so that in later decades of the nineteenth century, French bonbons lost their value to the candy that existed in a warm, happy, and distant past when people made candies "in the large iron pot, over a fire of glowing hickory coals, and made the occasion for many merry gatherings."[63] The oppositional nature of homemade candies to refined and effete bonbons marked the dark heavy molasses confections as those most suitably "old-fashioned" and hence more desirable at a time when refined sugar was so prevalent. A 1904 candy manual reiterated this opposition, printing recipes for "Old Fashioned Chocolate Caramels" and "Modern Chocolate Caramels" one after the other.[64] Not surprisingly, the presence of molasses distinguished the two: it appeared in the first recipe but was replaced with brown sugar in the second.

By century's end, molasses signified not only a temporal past but also a geographical past. The South, whose slaves produced and consumed most of the stuff, also represented the backward and unmodern. The following typified the slave's diet: "Molasses was eaten mixed with the fat obtained when fat salt pork was fried. This mixture, with corn bread and fried pork, formed the basis of the regular diet for each meal every day in the year. . . . Hot water sweetened with molasses was the beverage used with the meals."[65] A dark, viscous substance, molasses countered the granulated, pure, white refined sugar of the genteel North. It symbolized things quaint and distant, hearkening back to a former time and another place—a racist and exploitative past still thriving in the Reconstruction South. Contradictorily, that people associated molasses with quaintness and even poverty concealed the fact that by the end of the century, it rivaled refined sugar in price. In 1872, sugar cost 12 cents a pound on average, while molasses cost 70 cents a gallon (or .0875 cents a pound). By 1902, sugar sold on average for 5 cents a pound, and molasses for 49 cents a gallon (or a little more than 6 cents a pound).[66] Refining cane and beet sugar had become so efficient that it left little by-product. The technological developments that decreased the cost of and modernized refined sugar also elevated the prices of molasses and brown sugar. As Sidney Mintz has remarked, "the traditional sugars survive as heirlooms of a sort—expensive relics of the past—whereupon they may reappear as stylish 'natural' or conspicuous items on the tables of the rich, whose consumption habits made them rare and expensive in the first place."[67] People looked fondly back at an unmechanized past and paid more for goods that seemed to embody it.

Old-fashionedness applied not just to ingredients but to candy-centered activities as well, like the safe and wholesome taffy pull. A home-centered activity, the taffy pull gained popularity during the era that also celebrated the Colonial Revival, a response to America's increasing anxiety over new technologies and new immigrants populating the country. Retreating to the past helped Americans feel safe in the face of surrounding unfamiliarity. The author of *Dr. Miles' Candy Book* asked, "Who, indeed, will be bold enough to decry the old fashioned 'candy pull' or to assert that the eating of even a goodly portion of 'taffy' is injurious to humanity?"[68] Taffy pulls also enabled children to make things that they could give as gifts, thereby developing social skills: "They . . . will take great delight in being able to present to parents or friends a box of delicious bonbons made by themselves," wrote another author.[69] Candy-making parties for children created a future-oriented nostalgia that was orchestrated by mothers who wanted to construct fond memories for their children to experience as adults.

Ornamental Desserts

Homemade candies meant for one's intimate circle of family and close friends could be imperfect and were often valued precisely for their imperfections. However, homemade desserts served in more formal settings met with much closer scrutiny by the people eating them. Fancy mousses, cakes, puddings, soufflés, and blanc-manges graced formal dinner tables as the meal's crowning achievement. Desserts made personally by the hostess reflected and displayed the status of the family. Influenced to a large extent by the fashions of interior design, a dessert's appearance manifested in yet another way the aesthetics of showmanship. Ice creams from copper molds became lambs, flowers, thistles, and dogs reminiscent of family crests. Marzipan paste and spun sugar formed centerpieces that looked like fountains and birds' nests, mimicking the natural themes decorating wallpaper and upholstery. Elaborate gelatin forms, glistening and studded with all manner of fruits, resembled the parlor aquarium or terrarium that revealed yet concealed internal treasures. And at the same time that they mimicked other contemporary signifiers of high fashion, fancy desserts also referred to the grand sugar sculptures, ice towers, and fruit pyramids of royal banquets.

Even though a middle-class woman's lack of real work, often determined by the number of servants in her employ, defined her social standing, the inverse principle held true when it came to making desserts, which was an activity not usually relegated to a woman's staff. Homemade desserts evinced a family's ability to afford the costly but necessary raw materials and tools and embodied a housewife's amount of expendable time and energy to devote to such seeming frivolities of fashion. Gilded Age American culture expected a genteel woman to be well versed in household matters from cooking to cleaning only so that, as a skilled domestic manager, she could more deftly choreograph the tasks of her servants. Although she calculated the household budgets and compiled dinner menus and shopping lists, the woman of the middle-class house was not expected to "get her hands dirty" by performing the actual jobs of shopping and cooking. Her social standing resided in her ability to be able to pay someone else to do pedestrian tasks, proof of her elevated nature. However, when it came to sweets—including candies and baked goods—a woman's pride, status, and honor (and by extension, that of her family) rested in making such things. A woman's reputation within her social circle rested on her successful and impressive desserts.[70]

The Art of Confectionery, published in 1866, offered advice for a woman making her own confections and acknowledged the class-based division of labor common in all elite households, stating, "While the preparation of soups, joints, and gravies, is left to ruder and stronger hands, the delicate fingers of the ladies of the household are best fitted to mingle the proportions of exquisite desserts, to mould the frosted sugar into quaint and fanciful forms, and to tinge these delicious trifles with artistically arranged colors. It is absolutely necessary to the economy of the household that this art should form a part of every lady's education." Like needlework and piano playing, the art of confectionery counted as a requisite skill for a "lady" that both added to her overall refinement and sustained the gentility of her home. The above author continued, "This fact is becoming generally acknowledged, and the composition of delicate confections is passing from the hands of unskilled domestics into the business and amusement of the mistress of the household."[71] Even authors of various confectionery manuals connected objects and persons materially: "delicate fingers" made "exquisite desserts."

Although candies and desserts attracted the interest of one's female guests in particular, whose own abilities they judged comparatively, people spoke of them as "trifles" and "dainties," seemingly inconsequential accompaniments to a dinner. Judging each other on dessert fashions was a conceit enjoyed by the middle and upper classes. The ability to create successful and original desserts paralleled the mastery of other "feminine" skills, like making whimsies or practicing flower arranging. Labeled "amusements," the outward trivialization of these activities simultaneously elevated the pastimes to the status of a "genteel art."

Outside influences clearly affected how women behaved in their homes. Cooking classes, commercial expositions, window displays, and advertising informed them about the current fashions and which ones they should try implementing at home. The *Boston Cooking School Magazine,* for example, frequently published questions from housewives that reveal the extent to which the public sphere influenced them. One woman asked, "Will you give me, through the columns of your magazine, a recipe for Marshmallow Chocolate Cake? I purchased at the Woman's Industrial Union a sheet of this cake, which was so delicate that we were delighted with it."[72] Another woman wanted a recipe for macaroons; not generic ones, but "*just like* those we buy of the baker."[73]

Other women's magazines, including the *Ladies' Home Journal,* also shaped the tastes and diets of the middle-class American home at the turn of the century. They frequently published suggested menus for the entire week with accompanying recipes. These menus, heavy on meats, starches, and sugar, influenced the American diet and included such desserts as "Caramel Custard with Caramel Sauce," "Cream Cakes," "Prunes Moulded in Lemon Jelly with Whipped Cream," "Tapioca Cream," "Snow Pudding," and "Charlotte Russe, with French Fruit," turning America's culinary habits into one big dessert culture. In fact, 42 percent of the recipes appearing in the *Ladies' Home Journal* from 1884 through 1912, or 2,086 of the 4,942 total, were for desserts; the ratio was even higher in the manuscript recipe books of the same time.[74]

Women's periodicals indirectly promoted the sugar interests in other ways, too. By providing direct advertising from candy and sugar companies, they tacitly endorsed a company's products. "A London dentist," read one advice column, "holds that . . . sugar is nourishing to the teeth rather than otherwise."[75] Another author opined, "Americans are proverbially fond of sweets, and, in many families, the dessert is an item requiring much attention."[76] Readers, swayed by these trusted authors who made sugar consumption permissible, could not help but be affected by the sometimes subtle and sometimes overt inducements to eat it. By 1900, middle-class families ate so many desserts that one writer claimed, "We got so tired of the stated desserts, and had fallen so lamentably into a routine of the same pies and puddings, that the family struck one night."[77]

More than popular literature, developing technologies in kitchen gadgets and food itself helped bring the artistic forms of fancy desserts into poorer homes. These innovations blurred the distinctions between upper- and lower-class status foods and the public and private spheres. Machine-made, packaged instant foods helped even the most unskilled women not educated to such things to make fancier desserts that mimicked upper-class dishes. In addition, kitchen innovations also enabled the production of fancier desserts in more predictable and reliable ways. In the case of candies, this came in the form of the candy thermometer, which meant that women no longer needed to test boiled sugar by hand, but could gauge its temperature to the exact degree through a more "scientific" means. Like the processes that helped mass produce the decorations for wedding cakes, for example, these domestic technological developments democratized what were once exclusively upper-class goods. They blurred the aesthetic, material lines that had formerly distinguished classes and taste cultures from one another and brought the masculinized world of technological rationality into the feminized world of domesticity.

Many food companies increased familiarity with and popularity of their new products by disseminating advertising recipe booklets, which promised the successful creation of formerly exclusive desserts. Producers printed these booklets by the thousands and tied them to or inserted them into everything from apple peelers to boxes of macaroni. The sheer volume of such printed ephemera quickly rendered manuscript recipe books obsolete. As women turned to new food technologies, they necessarily had to use company-sponsored recipe booklets for instructions on how to use these new foods. Therefore, the personal manuscript recipe books that women used to keep—which through their circulation, solidified community bonds and reinforced family traditions—were displaced by a new kind of culinary heritage created in company test kitchens and disseminated to all.[78] The desire for both convenience foods and status items meant that women radically altered some of their traditional foodways and replaced others with new foods, new food technologies, new recipes, and new ways to remember these recipes.

A clear example of how advertising succeeded in getting people to replace old with new can be seen in the success of granulated gelatins, which arrived on the market during the late 1880s. For centuries, women had made their own gelatins for aspics, jellies, pies, and puddings by boiling calves' feet in order to release the collagen, which has gelling properties.[79] The entire process took hours, finally rendering a yellowish, earthy-smelling gelatinous mass.[80] The advent of a crystallized gelatin in powdered form revolutionized many aspects of cookery, making it much easier to obtain fine-consistencied gelatins that people found more aesthetically pleasing and palatable. The use of instant gelatins also meant that formerly upper-class dishes, like transparent fruited gelatin desserts, appeared at the tables of the middling and lower classes. Recipe booklets and flyers both advertised the new products and taught people how to use them. Companies such as Plymouth Rock, Junket, Crystal, Cox, Knox, and Jell-O published multicolored booklets and pamphlets that touted the purity, convenience and economy of their products (traits lacking in traditional calf's foot jelly) and provided innovative recipes for new dishes. More significantly, the booklets, often featuring full-color pictures of desserts formerly found only in upper-class cookbooks, articulated the promise of status and fancy goods in their visual and textual rhetoric and brought it into the homes of the less fortunate.

One of Jell-O's earliest flyers, from 1902, included recipes for such exotic desserts as "Banana Cream," "Cherry Soufflé," and "Jell-O Snow Pudding" (in addition to some savory dishes, such as "Jell-O with Roast Fowl," and a "Shredded Wheat Jell-O Apple Sandwich"). Jell-O recognized women's self-consciousness regarding reputation and class by mentioning that the product was "Used by the finest hotels and restaurants throughout the country." The flyer questioned and promised, "Why use Gelatine and spend time soaking, sweetening and flavoring, when JELL-O produces exactly the same results. No trouble and less expense. No dessert more attractive. No dessert half so good. Simply add hot water and set away to cool. That's all. It's perfection. A sure surprise to the housewife."[81] And a surprise it was—it offered and delivered what many lower- and middle-class housewives sought: a pure, quick, easy, attractive solution for dessert. Jell-O manufactured not only a convenient product but also one that placed fancy-looking desserts within the reach of those formerly deprived of them. Initially, one could enjoy Lemon, Orange, Strawberry, and Raspberry at a fairly affordable 10 cents a box; two years later Jell-O added Chocolate and Cherry. Used together, Jell-O flavors could produce "an

FIGURE 33. High aspirations were possible with Jell-O. Jell-O centerfold. Genesee Pure Food Co., *Jell-O: The Dainty Dessert* (Leroy, N.Y.: Genesee Pure Food Co., 1905), 8–9. Collection of the author.

unlimited number of combinations," thus challenging the notion that mass production meant uniformity.

From their earliest publication, Jell-O's multiple-page booklets used full-color covers and centerfolds to complement their cornucopia of recipes and provide illustrative proof of their feasibility (fig. 33). In their pages, an ordinary woman found out how to make the formerly inaccessible desserts, now transmuted by technological convenience: "Ambrosia Jell-O," "Pineapple Trifle," and "Charlotte Russe," among others, could grace the table of her family. In 1904, two years after its mass introduction, Jell-O advertisements continued to explain the substance itself, "a combination of gelatine and choice flavors, sweetened and prepared in such a manner that the article is always uniform, and a delicate dessert can be made in a very short time simply by adding hot

FIGURE 34. The Knox Cherubs. Chas. B. Knox Co., *Thirteen Reasons for Using Knox Gelatine* (New York: Chas. B. Knox Co., 1910). Collection of the author.

water and setting away to cool." Still a relatively novel product, Jell-O's advertising rhetoric also reassured women about safety, describing it as a "strictly pure and wholesome food," "approved by the Pure Food Commissioner."[82] In addition, the company's promotions reinforced its true purpose by playing the gentility card with such statements as, "All good house-keepers have certain desserts and dishes which they take special pride in making and serving, and the desserts prepared from recipes in this booklet will be sure to please the eye and taste of the most fastidious."[83]

Jell-O and other companies embraced the liberating idea of modernity and projected it back to women in the form of lustrous worlds populated with brilliantly shimmering dishes that lay within their ability to make. They portrayed simplicity in a number of ways that incorporated now familiar tropes associating sweetness with the feminine world. For example, the Jell-O Girl embodied inner goodness, innocence, and a youthfulness that when associated with the product meant purity and ease of preparation. Other gelatin companies employed similar strategies: Junket also had a girl, and Knox used angelic cherubs (fig. 34); Plymouth Rock capitalized on the nostalgic simplicity revered by the Colonial Revival through its name and trademark Pilgrim woman (fig. 35).

FIGURE 35. Reassurance about the new by association with something old was Plymouth Rock's advertising strategy, capitalizing on nostalgic trends like the Colonial Revival. The Plymouth Rock Gelatine woman. Plymouth Rock Gelatine, *Dainties and Household Helps* (Boston: Plymouth Rock Gelatine, ca. 1910), front cover. Collection of the author.

Knox's advertising strategy, equally sophisticated, complemented Jell-O's. As a popular and keen competitor of Jell-O, Knox integrated many of the former's advertising techniques yet emphasized cleanliness and the convalescent properties of its product over novelty and variation. An 1896 recipe booklet stated that its Calves Foot Gelatine ("The Purest Made"), "is recognized to-day as the Standard by all users of pure food. It has no odor or taste to disguise, so requires less flavoring than any other; is clear and sparkling, needs no clarifying. You have, no doubt, noticed, while pouring the hot water on some gelatines, a sickening odor which will arise from it (this never will happen in pure gelatines), and shows that the stock is not pure, so is unfit for food."[84] While Jell-O touted its original recipes and advocated modernity by using bright and vibrant illustrations, Knox located itself in the past, stressing its connection to traditional calf's foot jelly, yet simultaneously sanitizing that past via mechanization, promoting no "sickening odors," which recalled the earth of its origins.

These new food products promised many things at once. They offered the possibility of endless and always interesting dessert variations, contradicting the machine and mass commodity ethic by suggesting that mass production could actually generate individuality. They also claimed, contradictorily, to be the one thing that would fit all tastes; whereas previous handmade goods occupied distinct market niches, so that some confections were marketed to and purchased by children, others by teens, and still others by older women, new dessert preparations acknowledged few, if any, discrete consuming groups. For example, Knox Gelatine, a universal palliative and treat, was the "one dessert for all appetites—for children as well as grown-ups."[85] *Dainty Junkets* contained recipes for baby food, children's desserts, and illness remedies, of "great value as a perfect healthfood especially for the young, growing child or as a nourishing, sustaining food for the sick and invalids."[86] Jell-O said it was "tempting for every member of the family, including the sick and convalescent."[87] So these new preparations could be, ideally, all things to all housewives, whether they needed a spur-of-the-moment impressive dessert or a nutritious meal for the sick.

While the food technologies and their advertising strategies may have been new, companies perpetuated well-established gender associations. They used images of women and, sometimes, children to represent purity, maintaining these consuming groups' affiliations with sweetness, delicacy, and desserts in

FIGURE 36. The Jell-O Girl as the literal embodiment of her product. Genesee Pure Food Co., *New Talks About Jell-O* (Leroy, N.Y.: Genesee Pure Food Co., 1918), front cover. Collection of the author.

general. Since color clarity and lack of smell indicated a pure gelatin, advertising emphasized them.[88] Boston's Crystal Gelatine claimed that it "sets very quickly, and makes a clear, transparent jelly, the old yellow color which nearly all jellies used to have being entirely eliminated." The cover of Crystal Gelatine's box, showing a woman holding up a mound of clear gelatin in front of her merely underscored the text: her head peeks through her gelatinous veil like one more piece of fruit embedded in it. A later Jell-O pamphlet made demolded Jell-O a part of the Jell-O girl herself—the folds of her dress suggest another Jell-O creation (fig. 36). Depicted this way, the product actually "embodies" her.

Yet while advertising may have exaggerated and entrenched these ideas, American culture itself continued to sustain the symbolism tying women to sweets and desserts. Entrepreneurial advertising strategies would never have been effective if they had not already had a strong cultural resonance. Advertising, especially when applied to dessert products like those at the turn of the century, succeeded in taking the desires and aspirations of the lower classes and combining them with already extant cultural beliefs and mores. Products like Jell-O presented magical, colorful, fantastic worlds that were a 10 cent purchase away. By populating their advertisements for "dainties" with images of women and children, companies also reinforced the associations that linked specific commodities with specific types of consumers.

Whether making candies or desserts, the audience addressed by home confectionery manuals and advertising recipe booklets really consisted of the middle classes, who knew what constituted upper-class aesthetics but did not have enough time to make things themselves. Women's magazines like the *Ladies' Home Journal* and the *Boston Cooking School Magazine* never pretended to be democratic, but rather concerned themselves with directing the middle-class home and maintaining very real if, at times, subtle class distinctions. Even suggestions for "Economical Menus" reflected this. A working-class immigrant family relied on a pot of stew and some bread for the day. A typical *Boston Cooking School Magazine* "economical" dinner for 1898 consisted of bisque soup, cold beef with horseradish, mashed potatoes, asparagus with drawn butter, string bean salad with French dressing, cabinet pudding with burnt cream sauce, and cereal coffee. As food technologies democratized sugar and desserts, the upper classes once again took refuge in elaborate, ornate homemade desserts that emphasized appearance. While practically anyone could eat Jell-

O, only a select few could enjoy gelatin desserts in the shape of a castle or a family crest. Although sugar was much cheaper by the turn of the century, the products made with it still varied in type, appearance, and quality. People continued to recognize these differences as they continued to use sugar as a potent social communicator in the guise of an innocent dietary flavoring.

Conclusion

∴ *The Sweet Surrender* ∴

Sugar's transformation into a feminine substance meant the erasure of its once masculine characteristics. This process, part of a larger nineteenth-century endeavor to maintain distinct gender roles, relied on a consensus of shared meanings largely brought about by institutional forces and collectively accepted by individuals. Conscious of the developing sensibilities of their consuming audiences, producers and advertisers created and perpetuated the distinct social lives for sweet things as they became democratized. We continue to live with the legacies they created. Sugar's proponents chose which historical associations to borrow, which to discard, and which new ones to introduce. And they enlisted a variety of media to disseminate these meanings to consumers, making sweetness a larger presence in the popular consciousness. Definitions of sweetness appeared in cookbooks and advice literature, medical treatises and popular periodicals, window displays and international exhibitions, and product packaging itself. A concomitant accretion of abstract qualities associated with sugar, under the general term of "sweetness," helped stabilize its meanings. This resulted in the wholesale emasculation of confections and their subsequent association with weakness, purity, and gentility—their feminization.

Why was sugar feminized? And why did Americans most identify with it via its appearance in and as confectionery? Unlike apparel, sugar did not outwardly make gender or class distinctions. Locked away in spice cabinets and apothecaries' shops, it did not initially have a proximal relationship to women, like domestic equipment, which was often gendered. Of course, women used sugar as an ingredient in cooking, a traditionally female task, but they also used other ingredients, like salt, which failed to accumulate similarly gendered meanings. No overtly functional or associative properties in sugar's early history account for its eventual feminization.

The forms of sugar that nineteenth-century Americans came to know as trifling confections began their history as substances with both practical purposes and weighty symbolic import. In the eighteenth century, producers and consumers alike perceived sugar as a masculine good, if they gendered it at all. European royalty of the Middle Ages appreciated sugar as much for its medicinal properties and its ability to make grand sculptures as they did for its novel taste. Recipes for sugary preparations that promised to cure anything from the Plague to simple ague appeared in early domestic recipe books and respected medical treatises. People initially valued chocolate, too, for its utilitarian properties. The Spaniards who colonized Mesoamerican cultures and crops found the beverage to be an effective digestive aid, in addition to believing in its powers as an aphrodisiac. People also imbued other sugar-laden substances with symbolic import, which eventually eclipsed their practical uses. The early taking of tea, chocolate, and coffee, for example, functioned as highly ceremonial occasions accessorized with elaborate cups, saucers, beverage pots, and utensils that highlighted and centralized sugar's important role as a status object. Likewise, towering ornamental sculptures appearing on the tables of royal banquets literally placed sugar at the center of important fêtes, designating both the ceremony and the host as powerful cultural entities.

Yet sugar's original masculinity existed not merely because of its practical uses and consumption contexts. Methods of production and distribution also defined sugar as a masculine good because these processes were direct manifestations of male political and economic power. In order to grow and cultivate sugarcane, men had to first conquer cultures and territory. In the fifteenth, sixteenth, seventeenth, and eighteenth centuries, the Spaniards, Portuguese, British, and Dutch took control of the Americas, the Caribbean, Jamaica, Barbados, the Canary Islands, and São Tomé, among other places. They enslaved

Africans and others to exploit these colonies. Great sailing vessels transported the proceeds to their own countries, where the most advanced contemporary technologies were used to process sugarcane into various grades of refined sugar, which was subsequently sold at great profit on domestic and foreign markets. These machinations involved the domination of people and nature at every interval. Making the sugar that Europeans and American colonists so blithely consumed meant the despoliation of lands and the deaths of human beings.

Only when refined sugar started losing its position as a status object did its properties come to mesh psychologically with nineteenth-century ideologies regarding the feminine sphere. Democratized to the point of insipidness, confectionery became increasingly linked to and discussed in terms of genteel feminine lives—ornamental, inessential, ephemeral, and easily dismissed. In addition, popular commentators resigned themselves to the fact that women in particular would eat sugar, being impractical by nature and little able to control their own desires: this superfluous commodity well suited their appetites.

Although American advertisers helped reinforce these meanings through the sheer quantity of promotional materials generated, consumers themselves, as autonomous individuals, initially chose to accept sugar into their lives and diets and eventually to use the collateral symbolism of confectionery to communicate with one another. Tailoring products to purchasers, confectioners relied on the sale of sugary things for their livelihoods, making fancy and imaginative treats to whet the appetites of purchasers young and old, rich and poor. The more astute businessmen knew how to make and display myriad kinds of candies and cakes in order to attract a broad clientele, from the girl with a penny in her pocket to the beau willing to buy a pound of fine French bonbons. Beet sugar production allowed Americans' desire for sugar to rise unabated, and confectioners transformed it from a raw, undifferentiated material into a culturally processed one. Sugar was no longer used to create concentrated markers of power but instead became important in a different way, because of its ubiquity and diffuseness. Casual observers remarked at the time that they saw an ice cream saloon or candy shop on every corner, that every child seemed to have something sweet in his or her mouth, and that women, insatiable, bought and ate pounds of bonbons.

While people took their cues from popular culture, they also integrated sugar into their lives in idiosyncratic ways, at times constructing elaborate rituals around its preparation and ingestion, often borrowing from European affectations. For example, ice cream making became an important middle-class family bonding activity in the second half of the nineteenth century, structuring many leisurely Sunday afternoons and becoming a ceremonial marker of special occasions. Many women tried their hands at making home-made candies and fancy desserts in their own homes, as much to impress their friends as to make sentimental gifts. Sugar in this context regained its status, but primarily in image-conscious female circles.

In addition, consuming sugar enabled people to interact with the commodity on levels that transcended the physical. Confections, a pleasure to eat, triggered fantasies in certain contexts. Children in search of penny candy "toys" spent hours in proto-capitalist behavior, browsing the glass display cases and jars full of saccharine treats, dreaming of abundance while trying to make an astute selection of candies for their pennies. For adults, consumption contexts like fancy ice cream saloons and soda fountains triggered similar fantasies of belonging to the genteel segment of society.

Institutional forces worked to mediate and channel singular experiences into a collective vocabulary of meaning. Individual types of confections had become highly differentiated by the time of their complete democratization, yet consumers agreed on their various ascribed meanings. Significantly, photography, chromolithography, and the burgeoning advertising and publishing industries transformed late nineteenth-century America into a visual culture that was highly invested in images and appearances. People quickly became adept at reading the meanings of these images, making pictures as effective communicators as written texts. The way consumers "read" two-dimensional images became the way they perceived three-dimensional ones as well, so that the very materiality of cultural artifacts held nuanced meanings for the people who bought and made use of them. As a result, the meanings of refined sugar became completely removed from anything having to do with its natural state. Rather, it became the basis for innumerable cultural products that people bestowed with various meanings, meaning-making processes that I have detailed in the preceding chapters. Sophisticated display and packaging strategies also helped to uproot the products of sugar from the realm of nature and submerge them into the world of culture. Confectioners' plate glass shopwindows and display cases objectified these goods, endowed them with magical properties, and made them ever more alluring by reason of their simultaneous visual accessibility and physical inaccessibility.

The eventual democratization of refined sugar and the confections made with it during the cultural climate of the nineteenth century accompanied a simultaneous psychological devaluation of these commodities and the people who consumed them. Forms of popular culture perpetuated a complex life history for sugar that both attached itself to and was determined by the people who ate it—women and children—making it unique among other inanimate goods of the time. The more people consumed confections, the less they were impressed by them. The more women relied on them as supplements or additions to their diets, the less men respected them—both women *and* confections. Beyond its incarnation as confectionery, sugar itself assumed a social life that transformed it from an inanimate state into something almost in possession of its own life force.

It might be easy for us today to dismiss the historical importance of sugar because of its successful cultural demotion barely over a hundred years ago. Yet in cultural context, the feminization of sugar reinforced the gender hierarchies that refined Americans tried so desperately to maintain in the late nineteenth century. Not only did sugar's feminization objectify women, but it also made them saccharine—nonessential, decorative, sweet, ethereal, and generally lacking in substance. This encouraged women's relegation to a separate sphere by associating them with the commodities they purchased. It also constituted a clearly articulated and naturalized way of seeing women as essentially different from and inferior to men. Since people increasingly defined themselves (and one another) vis-à-vis external, material goods, it was telling that women became associated with sugar and men with goods such as tobacco, alcohol, and meat in a kind of cultural shorthand. In addition, sugar's reputation alone determined that it was inessential, possibly unhealthy, and almost certainly addictive. It represented the perceived inability of women to control their own bodily urges and reaffirmed commonly circulating ideas that because of their "nature"—impracticality and lack of control and foresight—women needed tight reins to keep them in line. These remain important points to consider, because they tell us not only about consumers' relationships to a burgeoning industrial society but also how we arrived at the present-day meanings of many of the things we buy, consume, and give as gifts—whether we are conscious of them or not.

Postscript

The Sweet and Low Down

By taking refined sugar into our mouths, like nineteenth-century Americans, we accept its constellation of meanings into our lives. Sugar is so prevalent that we ignore it. The overwhelming and enduring success of sugar's infiltration into American culture is evident when one considers recent consumption statistics. Estimates of Americans' sweetener intake during the 1980s and 1990s range from between 136 and 165 pounds per person per year, with a 29 percent jump in consumption (up 35 pounds) between 1982 and 1998 alone.[1]

A change in the physical character of sweeteners, concurrent with this rise in consumption, has facilitated their ingestion and rendered them more invisible. Companies now favor alternative sweeteners to refined sugar, including saccharin, aspartame (otherwise known as Nutri-Sweet™), and high fructose corn syrup, which is much cheaper than refined sugar. Refined sugar has lost its former reputation as the new, pure, and modern sweetener, and has been displaced by noncaloric yet much sweeter sugar substitutes: saccharin is 300 times and aspartame 200 times sweeter than sugar.[2] Now outpaced by the noncaloric artificial sweeteners, refined sugar has become the honey and maple sugar of the twenty-first century.

The shift in taste preferences occurring since the 1970s indicates the ever-increasing role that manufacturers play in our dietary lives; most consumers buy their food and beverages ready-made and thus have little interest in or control over their constitutive ingredients. The shift in preferred sweeteners also reveals this turn-of-the-century's particular preoccupations. While nineteenth-century Americans signified their status through the possession of material goods, Americans of today display their status using their own bodies—bodies *as* material objects. The collective sweet tooth we acquired grows ever hungrier. But fulfilling such urges runs counter to our desire for trim figures, an omnipresent American obsession. Thus, many of us have abandoned refined sugar, becoming enamored instead of noncaloric sweeteners, ingested most often in liquid form, which advertisers assure us are "fat free."

Significantly, artificial sweeteners have expediently sated our conflicting desires for sweetness and thinness. In the 1950s, few people consumed artificial sweeteners. By 1978, each American annually ingested on average 7.1 pounds of saccharin (the only artificial sweetener on the market at the time). By 1984, saccharin consumption topped out at 10 pounds, and aspartame accounted for 5.8 pounds per person per year.[3] Most of this consumption occurred via diet sodas.

Figures for artificial sweeteners in carbonated beverages only account for part of the story. In addition to candy bars, ice cream, and other confections (a 32-ounce Mr. Misty Slush from Dairy Queen, for example, contains 28 teaspoons of sugar),[4] Americans increasingly eat sugar in foods not normally considered in need of sweetening. Of the twenty to thirty teaspoons of sugar we eat on average a day (about 100 to 128 pounds per year), half is concealed in "crackers, bottled salad dressings, soy sauce, packaged side dishes, peanut butter, and sweetened cereals," in addition to rice mixes, pizza dough, hotdogs, and myriad other products.[5] Ketchup, which the Reagan administration infamously decided could supply the vegetable component in school lunches, is 29 percent sugar.[6]

Nineteenth-century Americans habituated us to sweets of the processed variety; they also left us with a legacy of core cultural meanings related to sweet things, which we have adopted and adapted through time. The increasing promiscuity of sweeteners in our culture today is evidenced by the concrete statistics cited above. Yet refined sugar's progeny remain much more elusive and escape easy division into neat generic categories of meaning based on material form. They are dissolved into carbonated beverages and incorporated into bagels and canned soups. Most sweeteners enter our diets invisibly and largely unnoticed. Our palates have become more and more desensitized to them as our bodies become more and more desirous of them.

Lingering Symbols

The material elusiveness of popular sweeteners like high fructose corn syrup, saccharin, and aspartame corresponds to an infiltration of sweetness as a concept into our collective subconscious. Significantly, people still maintain many of the symbolic meanings established by the material categories discussed in the previous chapters, and our contemporary responses to and uses of confections would strike familiar chords in the nineteenth-century American psyche. Indeed, tracing the genealogy of such cultural meanings has been the main purpose of this book.

Adults still reward children with candy, a practice that nutritionists try to discourage.[7] Moreover, as in the nineteenth century, children are the key market for candy "novelties." Pairing candy and toys makes them even more attractive to kids: until very recently, toy "prizes" came in boxes of sugar-coated cereal. Plastic candy containers mimicking adult goods, reminiscent of penny candy premiums, also line store shelves and look like miniature versions of cellular phones, computers, and cars. Candies themselves often come in enticing shapes that make them look like jewelry or familiar action figures. Nostalgia has resurrected old candy forms, and at times the nineteenth-century responses to them. Candy cigarettes, for example, are still made by a few companies who dare to incur the wrath of tobacco opponents like Alan Blum, a physician who believes that chocolate cigarettes may encourage a real smoking habit. Edward Fenimore, founder of the Philadelphia Chewing Gum Company, stopped making bubblegum cigarettes in the early 1990s, explaining, "It's not moral to produce those kinds of things anymore."[8] Taking its cue from the nicotine "patch" used by reforming smokers, a recently developed vanilla-scented patch is supposed to cut the craving for sweets.

Heart-shaped boxes of soft candies still appear on grocery shelves around Valentine's Day and continue to be a popular romantic gift. But chocolates have become so common that even "fine" brands like the ubiquitous Godiva have lost their cachet, and gilt-foiled Ferrero Rocher are downright déclassé. In response, upscale shops are turning to bittersweet "boutique" chocolates, made with very little sugar, which resurrect the nineteenth-century idea of the fine chocolate. A 2000 news article reported that the exclusive Scharffen Berger chocolate, "made with restored vintage machinery and small-batch-roasted beans, reintroduced hand-crafting." One 9.7 ounce bar sells for $8.95, more than $14.00 a pound.[9] Artisan chocolates, produced by hand in small batches, are also increasing in popularity, perhaps in part because of recent movies like *Chocolat* and due to the "'imperfection' craze," which privileges evidence of the human hand or chic wear-and-tear in an era defined by computer-related technologies.[10]

In other cases, chocolates have merely become directives pointing to more glamorous products signifying the good life. For example, one recent advertisement in the *New Yorker* showed an open heart-shaped box of luscious bonbons fringed in red, with the word "Love" simply printed at the bottom. One chocolate nestled in among all the others is embossed with a Mercedes symbol: the chocolates are merely the vehicles for conveying the message of love as it is understood through the referent of Mercedes. The association of expensive cars, jewelry, and other luxury items with chocolates has become so ingrained that advertisers no longer even need to show the candy. "Melt Her Heart" a recent Valentine's Day ad in the Philadelphia *City Paper* advised, showing a gleaming diamond ring sitting in a fluted paper cup of the kind ordinarily reserved for chocolate bonbons. Here, even in its absence, fine confectionery evokes romance—chocolate's sensuous "melt in your mouth" quality melts the heart too.

Besides these lingering symbols, it may seem as if the chocolate entering the twenty-first century has lost much of its direct association with women's bodies. Yet a new advertising campaign for the Milky Way® Midnight™ chocolate bar immediately recalls the rhetoric of the Enrober. Made of dark chocolate that the company describes as "bold," the Midnight™ bar conjures up the danger and mystique of bare-breasted Aztecs and transgressive Spanish chocolate-eaters. One advertisement, appearing in a 2000 issue of *Spin* magazine, spotlighted the opened candy bar against a black background with the headline, "You don't unwrap it, You Undress it." Milky Way® invites the consumer to bare the "naked center" that the Enrober coated, yet keeps the gender implications intact. The chocolates are naturally female and highly sexualized, as the remainder of the advertisement's text makes undeniable: "Introducing Milky Way® Midnight™. Seductively bold chocolate, golden caramel and vanilla nougat. All covered up without any tricky hooks in the back."[11] Equally overt is Nestlé's 2001 advertising campaign for its "Treasures," which the company has suggestively nicknamed "The Tunnel of Love." Sanctioning the guilty pleasure of self-pleasure, the ad instructs, "Share them with someone special—you, for instance." Their phrase, "From you to you," stating outright a kind of gifting relationship proscribed in the nineteenth century, is even trademarked.

As the culmination of romantic love, wedding cakes continue to be requisite elements of wedding receptions, where they usually appear in their archetypal form—white, tiered, and decorated with flowers. And they still directly and unapologetically refer to the bride. One specialty cake maker admits, "A bride's gown is often my inspiration." Or she tries to coordinate the cake with other decorative elements involved in the ceremony, "creat[ing] frosting copies of the lace patterns around the edge of the bride's veil or the embroidery on the bodice of her dress."[12] But couples are also choosing more idiosyncratic versions of cakes that reflect both their willingness to incorporate familiar symbols into their rituals and their simultaneous need to accommodate personal preferences. One contemporary popular design consists of two individual tiered cakes (reminiscent of the separate bride's and groom's cakes) joined by a latticework bridge that also supports bride and groom figurines.

Straying even farther from tradition, one bride I know opted for a chocolate wedding cake with chocolate frosting because, simply, this was her favorite flavor. Her choice seemed acceptable to guests because the entire reception was "unusual," because the bride and groom also served Chinese food. The "unusualness" of the cake, then, fit the overall theme of the reception.[13] But even though the reception was "unusual," the couple did not forego the cake: it is still the case that the cake defines the ceremony. Without the cake, the wedding is incomplete.

More interesting, many people believe they are choosing novel wedding cake forms defying tradition when they are really revisiting artifacts from the past. Recalling the nineteenth-century practice of giving pre-boxed pieces of cake to guests to take home as souvenirs is a "big new trend . . . the tiny, single-serving wedding cake, sort of a fancy cupcake for brides," costing about $100 each and given to each guest, in addition to having a traditional wedding cake.

Wedding planners also consider novelty groom's cakes to be "untraditional" and decorate them with "something relating to the groom's life, his hobbies, his interests," like fish, computers, or musical instruments.[14] This kind of cake has not strayed far from the 1890 one embellished with the groom's favorite hunting dogs.

Modern culture's sardonic take on the wedding cake has a voluptuous woman jumping out of a giant cake, breaking free of her saccharine shell and offering herself up for masculine consumption. She is the anti-wedding cake, a particularly provocative culinary appropriation who mocks everything that the traditional version represents. She is sexual while the bride is chaste; she can liberate herself from the ornamental form while the bride, as caketop ornament, remains permanently immobilized. She is the postmodern version of the medieval subtlety.

Even domesticated sugar has experienced its own brand of resurgence. "Real" maple syrups, for example, while very expensive, continue to serve as "authentic" New England souvenirs, along with their mates, maple sugar candies, molded into maple leaf shapes. In an effort to materially reconnect with sweeteners at a time when we mostly consume them invisibly, manufacturers are producing packets of "sugar in the raw." Appearing at hip coffee shops in brown paper packaging, this semi-refined, dark crystallized sugar seems more wholesome and, strangely, untainted than its kin. As such, it carries its own brand of cachet valuable only because to us white refined sugar seems vulgar and sugar substitutes seem (and are) artificial. That honey, maple sugar, and brown (unrefined) sugars seem purer to us now than refined sugar is an interesting twist. As Claude Fischler points out, "It implies a vision of the world in which Nature is no longer viewed as untamed and menacing, but rather threatened by technology."[15]

In the form of confectionery and other incarnations, refined sugar continues to impinge on our lives in familiar ways. But our propensity to consume sweetness in liquid rather than solid forms also means it is that much more ephemeral. Sweets and sweeteners, substantive and meaningful, remain changing and elusive. We can see this clearly by how sweetness gets sprinkled into our spoken language in new ways that borrow heavily from past meanings.

⁓ *Sweet Talk* ⁓

"Sweet" as a descriptor has been used since at least the fourteenth century to characterize taste in addition to things "pleasant to the mind or feelings," or to things "not corrupt."[16] In early nineteenth-century America, people more often applied the quality of sweetness to smells: flowers and one's breath, optimally, were sweet. Prescriptive literature instructed women about achieving a sweet (meaning unblemished) complexion, while religious tracts focused on improving one's character to achieve a "sweetness" of soul. Not surprising, as refined sugar came more and more to occupy the American psyche, the idea of sweetness shifted to taste. The association of sugar with women and children was increasingly naturalized, and the terms for sweet foods became synonymous with certain types of people and their environs. Girls, made of "sugar and spice," grew up to be women who managed the "home sweet home."

Many of sweetness's abstract properties remain in current use, and Americans have added many more, reflecting just how lively and varied this realm of meanings has become—well beyond those grounded in material artifacts. Indeed, the lexicon of sweetness is a worthy subject of study in and of itself.[17] For example, many expressions retain lingering meanings of sweetness being pure, good, graceful, and somehow inherently right.[18] "Sweet" can also mean something excellent or simple. In the military, it is "weapons-speak for something workable," a usage that might suggest the necessary outward trivialization of otherwise very serious matters.[19] The "sweet spot" is the best place for a tennis racquet, baseball bat, or hockey stick to make contact with the ball or puck. One puts up more money to "sweeten the pot" in a poker game—to make the stakes higher and therefore the potential return for the winner more desirable; similarly, one adds more compliments to "sweeten someone up," or to "sweet talk" him or her in order to curry favor.

Love and romance are probably the most familiar contexts in which we employ the diction of sweetness. To be "sweet on" someone is to be enamored, and if this person feels the same way, he or she becomes one's "sweetheart," "sweetie," "sugar," "honey," "honey pie," or sometimes even, more extravagantly, "love muffin" (Sidney Mintz speculates that many of these appellations may recall traditional Easter themes—"the famed reproductive powers of chickens and rabbits, and the special significance of eggs"—that also appear as confectionery).[20] Within an amorous relationship, it is not out of line to request, "Give me some sugar," meaning, at the very least, a kiss from one's beloved.

The drug trade has also coopted the language of sweetness, a usage alluding not only to refined sugar's pleasurable and addictive aspects, but also to its chemical purity and granular quality, similar to many hard drugs. The terminology acknowledges both desire and need while suggesting that hard drugs parallel candy as treats for adults, "a modernised expression of the conflict between ethics and pleasure."[21] This association of drugs and sugar has inspired a constellation of related terms in modern parlance. The "candy man," a drug dealer, sells "sugar" and "candy"—heroin and cocaine. "Sweet Jesus" (not to be confused with the same expression as a declaration calling upon the exalted religious figure) is heroin, "sweet Lucy" is marijuana, "sweets" are amphetamines, and "sweet stuff" describes heroin or cocaine, which specialists particularize as "needle candy" and "nose candy."

Other phrases involving the word "candy" include three recent terms that describe different sensory experiences as insubstantial yet supposedly pleasing, but that are not truly so. "Ear candy" is a certain genre of banal, "easy-listening" music—music gutted of its original meaning and verve—often heard in elevators and while on hold on the telephone. It is a term meant for something innocuous, soothing, and generally lacking in content (hence the allusion to sugar) but is understood in common usage as something cloying and irritating. A piece of "eye candy" may be a particularly beautiful woman whose assets are purely physical in yet another saccharine triumph of surface over substance.[22] The newest incarnation of such language is "arm candy," a term for an attractive person used as a pleasing accessory at a public event, "implying a beautiful object to attach to your arm, for others to feast their eyes upon." One commentator described it as "a one-night nothing, a post-sexual image enhancer and ego booster. In the conventional pairing—powerful older man, stunning young woman—her presence assures the seething jealousy of other men and guarantees the intrigued interest of other women."[23] But a man who uses "arm candy" differs from a "sugar daddy," an older man used by a younger woman for his money. These phrases all carry clearly negative connotations whose derisiveness stems from the long-established objectification of women as sweet.

However, "sweet" is used in other contexts that have more subtle yet equally negative undercurrents based on sugar's tenacious and ubiquitous presence. People execute "sweet" deals, for example, with particular finesse and to the advantage of the one employing the word "sweet." A "sweetheart deal," a phrase more overtly pejorative, describes an arrangement that benefits all parties involved in its making by treating them with extra favor, but is illegal or unethical. It is also dubious to "sugarcoat" something—to present a situation in a better light, to make it more palatable. Likewise, labeling a benevolent gesture as "sweet" can also serve to discredit or devalue it in some way, drawing upon the domesticated and old-fashioned qualities of sweetness that suggest a quaintness, an innocence, a naïveté that the person employing the word "sweet" does not possess but confers upon the one who has performed the gesture "that was so sweet."

Because sugar's characterization as a feminine substance has become so ingrained, references to sweetness can be pejorative, especially when they jump the gender boundaries and describe men. For example, a "candy ass" is a weakling—a man who is effeminate. Similarly, a "candy kid" is a sissy and a "candy bar punk" is a young prisoner who has found protection with an older and dominant man. The masculinized world of sports, however, has embraced sweetness, frequently using it as a compliment that recaptures some of its earlier connotations that suggest something beyond immediate references to physical grace and athletic prowess. For example, the nickname of Walter Payton, the small but tough and talented Chicago Bears running back, was "Sweetness." "Sweet" Lou Whitaker played second base for the Detroit Tigers, a position that requires particular quickness and agility. Another quick and strong athlete who inhabits the pantheon of sweetness is Pernell "Sweet Pea" Whitaker, a successful welterweight boxer. And we cannot forget the similarly and suggestively monikered Sugar Ray Leonard and Sugar Ray Robinson. Qualities of sweetness that characterize certain athletes acknowledge, in part, their more feminine qualities: they not only possess strength and skill, but also are smaller, more agile, and demonstrate a sense of style that transcends the basic (and often brute) physical requirements of the game.[24]

While the material forms of sweetness may have changed since the nineteenth century, their distinctive objectiveness still conveys early meanings. We have merely adapted them to fit contemporary material dialects. As significant, the abstract associations derived from the qualities of sugar still exist in the linguistic realm. They are proof of sugar's ultimate internalization and how it finds external expression in contemporary American rhetoric that remains grounded, unconsciously, in its nineteenth-century origins and their myriad significations.

NOTES

Introduction

1. John Howard Payne's original song is titled "Home, Sweet Home" (1821), but the more familiar form without the comma is used throughout here.

2. *Pennsylvania Gazette,* 12 June 1766.

3. See esp. Edwin Wolf II, *The Book Culture of a Colonial American City: Philadelphia Books, Bookmen, and Booksellers* (Oxford: Clarendon Press, 1988), and John Carter Brown Library, *The Colonial Scene (1602–1800): A Catalogue of Books Exhibited at the John Carter Brown Library in the Spring of 1949* (Worcester, Mass.: American Antiquarian Society, 1950).

4. *Pennsylvania Gazette,* 15 February 1775.

5. See Karl Marx on the fetishism of the commodity in *Capital,* vol. 1 (1867; New York: Vintage Books, 1977), esp. 163–77.

6. Paul L. Vogt, *The Sugar Refining Industry in the United States: Its Development and Present Condition* (Philadelphia: University of Pennsylvania, 1908), 42.

7. The granulating machine dried the damp white sugar and reduced the size of the grain, making a finer product, which dissolved more easily. The vacuum pan heated the sugar by steam in an airtight compartment so that it boiled at a lower temperature. On cooling, "a rapid crystallization takes place, which produces that uniform fine grain, such as is required in Loaf Sugar" (Edwin T. Freedley, *Philadelphia and Its Manufactures* [Philadelphia: Edward Young, 1867], 470). The centrifugal machine consisted of small drums lined with mesh baskets, which revolved at a minimum of 1,600 times a minute. "The centrifugal force throws the molasses out through the wire network and leaves the sugar. After the syrup is driven out a little water is spurted into the drum and thrown out through the sugar. This washes out any syrup that may remain and also any remaining coloring matter" (Vogt, *Sugar Refining Industry,* 16).

8. John Heitman, *The Modernization of the Louisiana Sugar Industry, 1830–1910* (Baton Rouge: Louisiana State University Press, 1987), 56.

9. Neil Borden, *The Economic Effects of Advertising* (Chicago: Richard D. Irwin, 1942), 280.

10. Heitman, *Modernization,* 58.

11. Chauncey M. Depew, *One Hundred Years of American Commerce* (New York: D. O. Hayes, 1895), 626.

12. "Uncle Sam's Sugar Bowl," *Confectioner and Baker,* 15 December 1902, 2.

13. For more on consumers' ability to make their own meanings from the manufactured commodities around them, see Daniel Miller, *Material Culture and Mass Consumption* (New York: Blackwell, 1994), esp. ch. 10.

14. Depew, *One Hundred Years,* 625; Inter-University Consortium for Political and Social Research, Ann Arbor, Mich.

15. Vogt, *Sugar Refining Industry,* 1, 90.

16. On sugar in Great Britain, see Sidney Mintz, *Sweetness and Power: The Place of Sugar in Modern History* (New York: Penguin Books, 1985); on ice and ice cream in Europe and Asia, see Elizabeth David, *Harvest of the Cold Months: The Social History of Ice and Ices* (New York: Viking, 1995); on chocolate in Mesoamerica, see Sophie D. Coe and Michael D. Coe, *The True History of Chocolate* (New York: Thames & Hudson, 1996).

17. Thomas J. Schlereth, *Victorian America, 1876–1915* (New York: HarperPerennial, 1991), 157.

18. See Miller, *Material Culture and Mass Consumption.*

19. Victoria de Grazia, "Changing Consumption Regimes," in *The Sex of Things: Gender and Consumption in Historical Perspective,* ed. id., with Ellen Furlough (Berkeley: University of California Press, 1996), 19.

20. Arjun Appadurai, "Gastro-Politics in Hindu South Asia," *American Ethnologist* 8 (1981): 494.

21. John McCusker, *Rum and the American Revolution: The Rum Trade and the Balance of Payments of the Thirteen Continental Colonies* (New York: Garland, 1989), 58, 427–28.

22. For a fuller discussion of "habitus," see Pierre Bourdieu, *Distinction: A Social Critique of the Judgment of Taste* (Cambridge, Mass.: Harvard University Press, 1984); on semiotic systems and the creation of "myths," see Roland Barthes, *Mythologies* (New York: Hill & Wang, 1972).

23. See esp. Mintz, *Sweetness and Power,* 88–94, 124; Madeleine Pelner Cosman, *Fabulous Feasts: Medieval Cookery and Ceremony* (New York: George Braziller, 1976); and Terence Scully, "Mediaeval French *Entremets,*" *Petits Propos Culinaires* 17 (June 1984): 44–56.

24. Quoted in Philip P. Gott, *All About Candy and Chocolate: A Comprehensive Study of the Candy and Chocolate Industries* (Chicago: National Confectioners' Association of the United States, 1985), 151.

25. Marx, *Capital,* 1: 165.

ONE ⁖ *Sugarcoating History*

1. *Pennsylvania Gazette,* 13 January 1730.

2. Sidney Mintz, *Sweetness and Power: The Place of Sugar in Modern History* (New York: Penguin Books, 1987), 19.

3. Mary Hinman Abel, *Sugar as Food,* U.S. Department of Agriculture. Farmers' Bulletin No. 93 (Washington, D.C.: Government Printing Office, 1899), 10; Mintz, *Sweetness and Power,* 23.

4. Mintz, *Sweetness and Power,* 19–20, 23.

5. Elizabeth David, *Harvest of the Cold Months: The Social History of Ice and Ices* (New York: Viking, 1995), 228; Sophie D. Coe and Michael D. Coe, *The True History of Chocolate* (New York: Thames & Hudson, 1996), 50.

6. Mintz, *Sweetness and Power,* 23–98 passim.

7. Joop Witteveen, "Rose Sugar and Other Medieval Sweets," *Petit Propos Culinaires* 20 (1985): 24.

8. Mintz, *Sweetness and Power,* 90.

9. John Scoffern, *The Manufacture of Sugar, in the Colonies and at Home, Chemically Considered* (London: Longman, Brown, Green, & Longmans, 1849), 46.

10. Mintz, *Sweetness and Power,* 31–100 passim.

11. John McCusker, *Rum and the American Revolution: The Rum Trade and the Balance of Payments of the Thirteen Continental Colonies* (New York: Garland, 1989), 42.

12. Mintz, *Sweetness and Power,* 37.

13. Ibid., 39.

14. Richard Blome, *A Description of the Island of Jamaica* (London: J. B. for Dorman Newman, 1678), 30.

15. James Thorold Rogers, *A History of Agriculture and Prices in England,* vol. 6 (Oxford: Clarendon Press, 1887), 322, 445.

16. W. M., *The Queen's Closet Opened: Incomparable Secrets in Physick, Chyrurgery, Preserving and Candying, &c.* (London: Christ. Eccleston, 1662), 6, 52, 221, 263.

17. Mintz, *Sweetness and Power,* 53, 67, 39.

18. Anne Bezanson, Robert D. Gray, and Miriam Hussey, *Prices in Colonial Pennsylvania* (Philadelphia: University of Pennsylvania Press, 1935), 164–72, appendix, 360–88, table 1.

19. *American Weekly Mercury,* 29 December 1719.

20. *Pennsylvania Gazette,* 8 May 1729.

21. Ibid., 25 February 1736.

22. William Fox, *An Address to the People of Great Britain on the Propriety of Abstaining from West India Sugar and Rum* (London: Sold by M. Gurney, 1791), 2.

23. *No* RUM!*—No* SUGAR! *Or, the Voice of Blood, &c.* (London: Printed for L. Wayland, 1792), preface, 22.

24. William Cowper, *The Complete Poems of William Cowper* (New York: D. Appleton, 1843), 197.

25. "Free Groceries," *Pennsylvania Freeman,* 29 March 1838.

26. William Reed, *The History of Sugar and Sugar Yielding Plants* (London: Longmans, Green, 1866), 9.

27. Paul L. Vogt, *The Sugar Refining Industry in the United States* (Philadelphia: University of Pennsylvania, 1908), 6–8.

28. For example, some estimate that in 1816, a batch of refined sugar yielded only 50 percent refined sugar, another 25 percent in molasses, and the remaining 25 percent in "bastard sugar," whereas a hundred years later, refiners realized over 90 percent refined sugar (*A Century of Sugar Refining in the United States, 1816–1916* [New York: American Sugar Refining Co., 1916], 8–9).

29. Vogt, *Sugar Refining Industry,* 13–14.

30. Mintz, *Sweetness and Power,* 77–78.

31. Glen R. Conrad and Ray F. Lucas, *White Gold: A Brief History of the Louisiana Sugar Industry, 1795–1995* (Lafayette, La.: USWLA, 1995), 25.

32. See in particular Alan Taylor, *William Cooper's Town: Power and Persuasion on the Frontier of the Early American Republic* (New York: Knopf, 1995), ch. 5.

33. M. De Warville, "On Replacing the Sugar of the Cane by the Sugar of the Maple," *New York Magazine* 3 (1792): 484.

34. *A Statement of the Arts and Manufactures of the United States of America, for the Year 1810* (Philadelphia: A. Cornman, 1814), xxxvi, xxxvii.

35. "Cheap Substitute for Sugar," *New York Magazine* 6 (1795): 747–48.

36. Homer E. Socolofsky, "The Bittersweet Tale of Sorghum Sugar," *Kansas History* 16, 4 (1993–94): 276; Vogt, *Sugar Refining Industry*, 3.

37. See Lewis S. Ware, *A Study of Various Sources of Sugar* (Philadelphia: Henry Carey Baird, 1881).

38. Charles William Taussig, *Some Notes on Sugar and Molasses* (New York: William Taussig, 1940), 34; Vogt, *Sugar Refining Industry*, 58.

39. Edward Church, *Notice on the Beet Sugar. . . . Translated from the works of Dubrunfaut, De Domballe, and others*, 2d ed. (Northampton: J. H. Butler; Boston: Hilliard, Gray, 1837), 7.

40. Vogt, *Sugar Refining Industry*, 58.

41. Ibid., 59; Herbert Myrick, *The American Sugar Industry* (New York: Orange Judd, 1899), 31.

42. Myrick, *American Sugar Industry*, 7; Vogt, *Sugar Refining Industry*, 59.

TWO ∴ *Sweet Youth*

1. Quoted in *A Century of Sugar Refining in the United States, 1816–1916* (New York: American Sugar Refining Co., 1917), 6.

2. *Pennsylvania Gazette*, 29 September 1763.

3. Ibid., 9 August 1775.

4. Augustus Lannier inventory, New York City, 1811. Joseph Downs Collection, Winterthur Museum and Library, Winterthur, Delaware.

5. Chauncey M. Depew, *One Hundred Years of American Commerce* (New York: D. O. Haynes, 1895), 625.

6. Sebastian Henrion, receipt, 23 June 1826, Henry Francis du Pont Winterthur Museum and Library.

7. For specific figures, see Mathew Carey, *Appeal to the Wealthy of the Land* (Philadelphia: L. Johnson, 1833).

8. W. M., *A Queen's Delight* (London: Printed for Christ. Eccleston, 1662), 9, 220.

9. Louis Untermeyer, *A Century of Candymaking, 1847–1947: The Story of the Origin and Growth of the New England Confectionery Company* (Cambridge, Mass.: New England Confectionery Co., 1947), 10.

10. "Confectionary," *The Friend* 8 (1834): 141.

11. Ibid.

12. *Parley's Magazine* 3, 11 (1835): 310.

13. *The Colored American*, 8 April 1837.

14. Thomas Teetotal, "Henry Haycroft—A Story for Youth," *Temperance Advocate and Cold Water Magazine* 1 (1843): 38, 39, 40.

15. "Sweets for the Million," *Once a Week,* 12 March 1864, 320.

16. See, e.g., *Cyclopaedia of Useful Arts,* ed. Charles Tomlinson (New York: George Virtue, 1854), 2: 776–89.

17. Thomas Mills and Co., Business Ledger 1864–72, Hagley Museum, Wilmington, Delaware.

18. Ibid.

19. Thomas Mills and Co., Patent Labor Ledger, Mills Bros. Collection, Mercer Museum, Doylestown, Pennsylvania.

20. Thomas Mills and Co., Business Ledger; Thomas Mills and Co., Patent Labor Ledger.

21. *Supply World* 13, 5 (May 1900): 28.

22. William Leach, *Land of Desire: Merchants, Power, and the Rise of a New American Culture* (New York: Pantheon Books, 1993), 63.

23. Michael Barton, *Life by the Moving Road: An Illustrated History of Greater Harrisburg* (Woodland Hills, Calif.: Windsor Publications, 1983), 42–43.

24. Leach, *Land of Desire,* 63.

25. David Nasaw, *Children of the City At Work and At Play* (Garden City, N.Y.: Anchor Press, 1985), 117–18.

26. "The Census Bureau reported impressive aggregates: 756,000 children ages 10 through 15 'gainfully employed' in 1870; 1,750,000 in 1900; 1,990,000 in 1910 (subsequently re-estimated downward to 1,622,000)," David I. Macleod reports in *The Age of the Child: Children in America, 1890–1920* (New York: Twayne, 1998), 110.

27. David Nasaw, "Children and Commercial Culture: Moving Pictures in the Early Twentieth Century," *Small Worlds: Children and Adolescents in America, 1850–1950,* ed. Elliott West and Paula Petrik (Lawrence: University of Kansas Press, 1992), 18; see also Macleod, *Age of the Child,* who notes: "In 1896 only 7 percent of 2,000 California girls ages 7 to 16 received an allowance, whereas 69 percent sometimes earned money; 10 percent of boys enjoyed allowances, whereas 74 percent earned at least occasional money" (114).

28. Macleod, *Age of the Child,* 114.

29. Ibid., 111–14. See also *Children and Youth in America, a Documentary History, 1866–1932,* ed. Robert H. Bremner (Cambridge, Mass.: Harvard University Press, 1971), 634–41.

30. Nasaw, *Children of the City,* 144.

31. "A Curious Candy Store Boycott," *Confectioners' and Bakers' Gazette,* 10 February 1905, 25.

32. "Hints to Retailers," *Confectioners' and Bakers' Gazette,* 10 July 1902, 18.

33. "The Candy-Shop," *Arthur's Home Magazine* 31, 2 (1868): 320.

34. Nathaniel Hawthorne, *The House of the Seven Gables* (1851; New York: Bantam Books, 1981), 49.

35. Will S. Monroe, "The Money Sense of Children," *Pedagogical Seminary* 6 (1898–99): 152–58.

36. Nasaw, *Children of the City,* 118.

37. Henry S. Bunting, *Specialty Advertising: The New Way to Build a Business* (Chicago: Novelty News Press, 1910), 76; "Child Influence a Great Factor in Selecting Gift Media," *Novelty News* 10 (May 1910): 5.

38. Bunting, *Specialty Advertising,* 30–31.

39. Ibid., 27.

40. Richard Osman, "Simple, Inexpensive Toys Are Suitable and Satisfactory 'Personal Appeal' Mediums," *Novelty News* 9 (December 1909): 26. "The small boy with snakes, nails, stamps and 'things' in his pockets has a seemingly endless capacity for acquiring weird gadgetry. The girl of matching years may share much of this interest and be greatly moved by more feminine toys, games, dolls and appurtenances as well," George Meredith observes in *Effective Advertising with Premiums* (New York: McGraw-Hill, 1962), 88.

41. Thorstein Veblen, *The Theory of the Leisure Class: An Economic Study of Institutions* (1899; New York: Mentor Books, 1953), 121.

42. Bunting, *Specialty Advertising,* 76.

43. Herbert Palmer, *The Mistletoe Child: An Autobiography of Childhood* (London: J. M. Dent & Sons, 1935), 135.

44. This, too, recalls Hawthorne's *House of the Seven Gables.* The daguerreotypist Holgrave, discussing his own profession, could as easily be talking about desire and consumption when he says, "everything appears to lose its substance the instance one actually grapples with it" (31).

45. Susan Stewart, *On Longing: Narratives of the Miniature, the Gigantic, the Souvenir, the Collection* (Baltimore: Johns Hopkins University Press, 1984), 44.

46. Susan Willis, *A Primer for Daily Life* (New York: Routledge, 1991), 6.

47. "Penny Candies Up to Date," *Confectioners' and Bakers' Gazette,* 10 August 1905, 30.

48. Ibid.

49. John M. Miller & Son, "Christmas Goods for the Season of 1876," Mercer Museum, Doylestown, Pennsylvania.

50. *Confectioners' and Bakers' Gazette,* 10 November 1905, 13–14.

51. *The Youth's Companion,* 27 October 1892, 563.

52. Susan J. Terrio, *Crafting the Culture and History of French Chocolate* (Berkeley: University of California Press, 2000), 258.

53. "A Candy Box Automobile," *Confectioners' and Baker's Gazette,* 10 April 1905, 33.

54. Stewart, *On Longing,* 57.

55. "Decided Fancy Boxes Are Toys," *Confectioners' and Bakers' Gazette,* 10 November 1901, 31.

56. Karin Calvert, *Children in the House: The Material Culture of Early Childhood, 1600–1900* (Boston: Northeastern University Press, 1992), 113.

57. See Monroe, "Money Sense of Children," and "Earnings, Spendings, and Savings of School Children," *Charities* (20 June 1903): 602, a study that determined that gender differences also marked childhood consumption: more girls than boys saved their money; when spending their money, girls tended to purchase practical items (clothing, sewing supplies), while boys used theirs for immediate gratification and entertainment (toys, candy).

58. See Macleod, *Age of the Child,* 112–13.

59. Monroe, "Money Sense of Children," 154.

60. Interestingly, American dentists' connection of sugar with caries—something people *should* have been concerned about—did not occur until well into the twentieth century.

61. W. B. O'Shaughnessy, "Poisoned Confectionary," *Lancet* 2 (14 May 1830–31): 194; emphasis in original.

62. *The Art of Confectionery* (Boston: J. E. Tilton, 1866), 15.

63. *Supply World* 14, 6 (June 1901): 17.

64. Harvey Wiley, *Foods and Their Adulteration: Origin, Manufacture, and Composition of Food Products; Infants' and Invalids' Foods; Detection of Common Adulterations, and Food Standards* (1907; 2d ed., Philadelphia: P. Blakiston's Son, 1911), 486.

65. Frank A. De Puy, *The New Century Home Book* (New York: Eaton & Mains, 1900), 196.

66. Christine Frederick, *Household Engineering: Scientific Management in the Home* (Chicago: American School of Home Economics, 1923), 343; emphasis in original.

67. Bronson Alcott, *The Physiology of Marriage* (Boston: John P. Jewett, 1856), 41.

68. Nasaw, *Children of the City,* 131.

69. Barton, *Life by the Moving Road,* 104.

70. Sanford Bell, "An Introductory Study of the Psychology of Foods," *Pedagogical Seminary* 11 (1904): 68.

71. "Cigarettes to Minors," *Tobacco* 7 (7 June 1889): 4.

72. "Short Chats with Retail Dealers," *Tobacco* 7 (5 June 1889): 2.

73. Gordon L. Dillow, "Thank You for Not Smoking," *American Heritage* 32, 2 (February–March 1981): 94–107.

74. Patrick Porter, "Advertising in the Early Cigarette Industry: W. Duke, Sons & Company of Durham," *North Carolina History Review* 48, 1 (January 1971): 35.

75. Quoted in Richard L. Rapson, "The American Child as Seen by British Travelers, 1845–1935," *American Quarterly* 17 (1965): 521.

76. "The Young Thieves" (Philadelphia: American Sunday School Union, ca. 1860). Historical Society of Pennsylvania.

77. J. Adams Puffer, "Boys' Gangs," *Pedagogical Seminary* 12 (1905): 191–92.

78. Nasaw, *Children of the City,* 118.

79. "Candy Roulette Wheels Tempted Children," *Soda Fountain,* May 1906, 27.

80. "Pretty Girls Dice Shakers," *Confectioner and Baker,* June 1903, 4.

81. James Redfield, *Comparative Physiognomy or Resemblances Between Men and Animals* (New York: W. J. Widdleton, 1866), 270–71.

82. *Colored American,* 8 April 1837.

83. Hawthorne, *House of the Seven Gables,* 221, 227.

84. Marion Harland, *House and Home: A Complete Housewife's Guide* (Philadelphia: Clawson Brothers, 1889), 174.

85. Wiley, *Foods and Their Adulteration,* 486. Trade journals also tried to capitalize on these associations, suggesting not quite convincingly that a man who ate candy was a

temperate one: "If a young man brings you a box of candy, and sheepishly eats about half of it before you can eat six pieces, he is a young man most promising. You may be quite sure that he is not a cocktail young man, or one of the young men who think 'a little whiskey is good for the system.' Candy and whiskey rarely meet in young men's interiors" ("The Young Men Buying Candy," *International Confectioner,* January 1904, 28).

86. "High Balls in Candy Stores," *Confectioner and Baker,* April 1906, 23.

87. Sarah Tyson Rorer, "Why Sweets Are Not Good for Children," *Ladies' Home Journal,* March 1906, 38.

88. "Candy as a Cure and Substitute for Alcoholic and Similar Beverages," *Soda Fountain,* September 1907, 20.

89. Nasaw, *Children of the City,* 144.

90. *The New Cyclopedia of Domestic Economy,* ed. E. F. Ellet (Norwich, Conn.: Henry Bill, 1873), 527, 528.

91. "Recipes for Boys and Girls," *Boston Cooking School Magazine* 7 (1902–3): 460–61.

92. Janet M. Hill, "Candies for Boys and Girls to Make," *Boston Cooking School Magazine* 8 (1903–4): 272–74.

93. Marie Sault, "Decorative Functions for Midwinter," *Boston Cooking School Magazine* 1 (1896–97): 165–66.

94. Nelle M. Mustain, *Popular Amusements for In and Out of Doors* (Chicago[?]: L. A. Martin, ca. 1902), 331.

95. Carrie May Ashton, "Novel Fads in Children's Parties," *Boston Cooking School Magazine* 8 (1903–4): 197.

96. Harland, *House and Home,* 174.

THREE ∴ *Cold Comforts*

1. William Black, Diary. Virginia Historical Society, Richmond, Virginia. Photocopy in possession of the Hagley Museum and Library, Wilmington, Delaware.

2. Elizabeth David, *Harvest of the Cold Months: The Social History of Ice and Ices* (New York: Viking, 1994), 227–28.

3. Ibid., 26–27, 40, 67.

4. Ibid., 105.

5. Ibid., 118. David acknowledges that this description may be one more of popular lore than of fact.

6. Ibid., 310–11.

7. Hannah Glasse, *The Art of Cookery, Made Plain and Easy* (London: Printed for the author, 1758), 332.

8. "An Elegant Retreat," *Pennsylvania Gazette,* 30 September 1789.

9. J. B. Bordley, *Essays and Notes on Husbandry and Rural Affairs* (Philadelphia: Budd & Bartram, 1799), 370.

10. Thomas Moore, *Essay on the Most Eligible Construction of Ice-Houses* (Baltimore: Bonsal & Niles, 1803), 13, 26.

11. Richard O. Cummings, *The American Ice Harvests* (Berkeley: University of California Press, 1949), 2.

12. Quoted in Henry G. Pearson, "Frederic Tudor, Ice King," *Massachusetts Historical Society Proceedings* 65 (1933): 177–78.

13. Quoted in Harold D. Eberlein and Cortlandt Van Dyke Hubbard, "The American 'Vauxhall' of the Federal Era," *Pennsylvania Magazine of History and Biography* 68 (1944): 168, 164.

14. "Philadelphia Vauxhall," *Aurora Daily Advertiser* April 1816.

15. Eberlein and Hubbard, "American 'Vauxhall,'" 170.

16. "Philadelphia Vauxhall."

17. Quoted in Barbara Wells Sarudy, "Genteel and Necessary Amusements: Public Pleasure Gardens in Eighteenth-Century Maryland," *Journal of Garden History* 9, 3 (1989): 120.

18. *The Tourist's Guide Through the Empire State,* ed. S. S. Colt (Albany, N.Y.: Mrs. S. S. Colt, 1871), 4 (emphasis in original), 121.

19. Joseph Jackson, *The Encyclopedia of Philadelphia* (Harrisburg, Pa.: National Historical Association, 1933), 1155–56.

20. Lillian Foster, *Way-Side Glimpses, North and South* (New York: Rudd & Carleton, 1860), 15–16; Junius Henri Browne, *The Great Metropolis; A Mirror of New York* (Hartford, Conn.: American Publishing Co., 1869).

21. "Ice Creams," *Poulson's Daily Advertiser,* 29 June 1809.

22. *Poulson's Daily Advertiser,* 1 July 1809; emphasis in original.

23. "The Progress of Luxury," *Saturday Courier,* 27 June 1835; emphasis in original.

24. "Masser's Self-acting Patent Ice-Cream Freezer and Beater," *Godey's Magazine and Lady's Book,* July 1850, 124.

25. M. Emy, *L'art de bien faire les glaces d'office, ou les Vrais principes pour congeler tous les rafraîchissemens* (Paris: Le Clerc, 1768).

26. "Parkinson's Refreshment Saloon," *Sunday Dispatch,* 2 September 1855.

27. "Parkinson's Restaurant," *Gleason's Pictorial Drawing-Room Companion,* 28 May 1853, 344.

28. "Progress of Luxury"; emphasis in original.

29. "Parkinson's Restaurant," 344.

30. "Progress of Luxury"; emphasis in original.

31. Isaac Rodgers receipt, Library Company of Philadelphia, 26 December 1845.

32. "Ice Cream," Untitled newspaper clipping, Poulson's Scrapbooks, Library Company of Philadelphia, 3 July 1847.

33. George Foster, *New York in Slices* (New York: W. F. Burgess, 1850), 72.

34. Solon Robinson, *Hot Corn: Life Scenes in New York* (New York: Dewitt & Davenport, 1854), 220–21. I would like to thank Paul Erickson for bringing this and other city mysteries to my attention.

35. "Ice Cream Saloon," *Media Advertiser,* 4 July 1855.

36. "A Pleasant Retreat," *Delaware County Republican,* 2 July 1858.

37. Ned Buntline, "The Death-Mystery: A Crimson Tale of Life in New York," *New York Mercury,* June 1861.

38. "Mrs. Balduff's Ice Cream Saloon and Confectionery Store," *Delaware County American,* 2 August 1865.

39. "The Presidential Progress," *New York Herald,* 20 February 1861.

40. "The Festivities of Mrs. Lincoln," *Charleston Mercury,* 5 April 1862.

41. "The Ice Question," *Charleston Mercury,* 20 March 1862.

42. "New Opera House," *New York Herald,* 26 April 1862.

43. "The Close of the Summer Season," *New York Herald,* 2 September 1863.

44. *Delaware County American,* 2 August 1865.

45. "Come One! Come All!" *Delaware County Republican,* 22 July 1864.

46. "Mrs. Hall's Ice Cream Saloon," *Delaware County American,* 4 July 1866.

47. "Rice's Bakery and Confectionery," *Delaware County American,* 23 April 1866.

48. *Delaware County American,* 9 June 1869.

49. "Ice Cream Manufactured by Steam," *Delaware County American,* 23 June 1869.

50. I thank Harold Screen for his input and generous access to sources regarding the history and development of the modern soda fountain.

51. See, e.g., Paul Dickson, *The Great American Ice Cream Book* (New York: Atheneum, 1972).

52. R. R. Shuman, "Psychology of the Soda Fountain," *Soda Fountain,* April 1907, 24.

53. "'Soda' Water Styles," *Carbonated Drinks,* January 1878, 29.

54. Charles Lippincott and Co., *Catalogue of Apparatus* (Philadelphia: Charles Lippincott, 1876), 111, 120.

55. *The American Dispensers Book* (Boston: American Soda Fountain Co., 1903), 11.

56. "'Soda' Water Styles," 29.

57. Shuman, "Psychology," 24.

58. A. Emil Hiss, *The Standard Manual of Soda and Other Beverages* (Chicago: G. P. Engelhard, 1897), 24.

59. Shuman, "Psychology," 23.

60. "How to Run a Soda Fountain Successfully," *Confectioner and Baker,* May 1903, 21.

61. *American Dispensers Book,* 18.

62. Shuman, "Psychology," 24.

63. "Cool Store Attracts Trade," *Soda Fountain,* June 1906, 38.

64. Hiss, *Standard Manual,* 24.

65. "Some Soda Water Statistics," *Scientific American,* 12 August 1899, 99.

66. "College Girls Win Strike Against the 10-Cent Soda," *Soda Fountain,* March 1908, 20.

67. Susan Porter Bensen, *Counter Cultures: Saleswomen, Managers, and Customers in American Department Stores, 1890–1940* (Urbana: University of Illinois Press, 1988), 20. For more on women's relationships with department stores, see William Leach, *Land of Desire: Merchants, Power, and the Rise of a New American Culture* (New York: Pantheon Books, 1993).

68. E. C. Tracey, "Soda Invades Whiskey Abode," *Soda Fountain,* February 1906, 12.

69. "Fountains as an Aid to Temperance," *Soda Fountain,* December 1908, 20–21, 63.

70. "No More 'High Balls' in Detroit," *Soda Fountain,* May 1908, 20.

71. "Fountain and Cigar Store United," *Soda Fountain,* December 1908, 30.

72. "The Only Combination Tobacco-Store Soda Fountain Store in New Orleans," *Soda Fountain,* February 1909, 31.

73. "Men in Chicago Like Sweet Soda," *Soda Fountain,* May 1906, 6.

74. "Is the Sweet Tooth Really Better Developed in Men Than in Women?" *Soda Fountain,* September 1909, 29.

75. Perkins scrapbooks, Historical Society of Pennsylvania, vol. 80.

76. Quoted in Jeanne H. Watson, "Cult of Domesticity," *Nineteenth Century* 7, 4 (Winter–Spring 1982): 39.

77. "Successful Methods Applied to Soda Fountain in Chicago Ghetto," *Soda Fountain,* September 1908, 33–34.

78. George Rice, "Study of the Hands That Grasp the Glasses at Busy Fountains," *Soda Fountain,* February 1907, 15.

79. "Ice Cream," Untitled newspaper clipping, Poulson's Scrapbooks, Library Company of Philadelphia, 3 July 1847.

80. *City Cries* (Philadelphia: George S. Appleton, 1850), 65.

81. Daniel M. Bluestone, "'The Pushcart Evil': Peddlers, Merchants, and New York City's Streets, 1890–1940," *Journal of Urban History* 18, 1 (November 1991): 74.

82. "Ice Cream," *Sunday Dispatch,* 2 September 1855.

83. C. H. King, "Ice Cream Poisoning," *Bakers' Helper,* 1900, 378.

84. Anne Cooper Funderburg, *Chocolate, Strawberry, and Vanilla: A History of American Ice Cream* (Bowling Green, Ohio: Bowling Green State University Popular Press, 1995), 75.

85. See David, *Harvest of the Cold Months,* esp. 1–40, 141–72.

86. "Dangers of Peddled Ice Cream," *Confectioner and Baker,* June 1902, 15.

87. "Ice Cream Sandwich Machine," *Bakers Review,* 15 April 1904.

88. "The itinerant vender. . ." *Confectioners' and Bakers' Gazette,* 10 November 1902, 16.

89. *Soda Fountain,* June 1906, 32.

90. "Against Street Vendors," *Confectioner and Baker,* April 1906, 19.

91. Harvey A. Levenstein, *Revolution at the Table: The Transformation of the American Diet* (New York: Oxford University Press, 1988), 40.

92. Louis N. Magargee, *Seen and Heard* 4, 207 (21 December 1904): 4979.

93. "Ice Cream Makers Discuss Soda Fountains," *Soda Fountain,* November 1909, 46.

94. "New Ice Cream Fields," *International Confectioner,* May 1908, 27.

FOUR ⁓ *Sinfully Sweet*

1. Marion Harland, *House and Home: A Complete Housewife's Guide* (Philadelphia: Clawson Brothers, 1889), 174.

2. Sophie D. Coe and Michael Coe, *The True History of Chocolate* (New York: Thames & Hudson, 1996), 37–39, 40, 43–46.

3. Ibid., 108.

4. Ibid., 71, 106–7, 108.

5. Ibid., 23, 24–25.

6. See ibid., esp. 17–34, for more detailed technical information.

7. Ibid., 130–31.

8. Ibid., 114–15.

9. Ibid., 156, 142, 169.

10. Richard Blome, *A Description of the Island of Jamaica* (London: J. B. for Dorman Newman, 1678), 4–5.

11. Quoted in Coe and Coe, *True History,* 175.

12. Louis Lémery, *Traité des aliments* (1702), trans. under the title *A Treatise of All Sorts of Foods, Both Animal and Vegetable: Also of Drinkables* (London: T. Osborne, 1744), 364.

13. Quoted in A. Saint-Arroman, *Coffee, Tea and Chocolate: Their Influence upon the Health, the Intellect, and the Moral Nature of Man* (Philadelphia: Townsend Ward, 1846), 3.

14. Quoted in Coe and Coe, *True History,* 159.

15. "Chocolate," *Saturday Evening Post,* 9 July 1859, 5.

16. Coe and Coe, *True History,* 176.

17. Samuel Sewall, *The Diary of Samuel Sewall* (New York: Farrar, Straus & Giroux, 1973), 380, 476, 563, 563–64, 570.

18. Coe and Coe, *True History,* 230.

19. See, e.g., William Rawle, *Pennsylvania Gazette,* 4 December 1729, and Evan Morgan, *Pennsylvania Gazette,* 13 January 1730.

20. Coe and Coe, *True History,* 230.

21. Lémery, *Treatise,* 365.

22. "Cacas—Theobroma; Or, the Chocolate Tree," *American Farmer,* 2 August 1822, 145.

23. "Holcus bicolor," *Hazard's Register of Pennsylvania* 4, 12 (19 September 1829): 208.

24. "Chocolate," *Scientific American* 8 (1852): 3.

25. Nicholas Thomas, *Entangled Objects: Exchange, Material Culture, and Colonialism in the Pacific* (Cambridge, Mass.: Harvard University Press, 1991), 27–30.

26. Coe and Coe, *True History,* 251.

27. Susan J. Terrio, *Crafting the Culture and History of French Chocolate* (Berkeley: University of California Press, 2000), 9, 23.

28. "New Year's Day in Paris," *Album and Ladies' Weekly Gazette,* 19 July 1826, 3.

29. Charles Elmé Francatelli, *French Cookery. The Modern Cook: A Practical Guide to the Culinary Art in All Its Branches* (Philadelphia: Lea & Blanchard, 1846), iii, vi; emphasis in original.

30. Wolfgang Schivelbusch, *Tastes of Paradise: A Social History of Spices, Stimulants, and Intoxicants* (New York: Pantheon Books, 1992), 91.

31. S. Henrion, Receipts, Henry Francis Du Pont Winterthur Museum and Library. Mathew Carey, *Appeal to the Wealthy of the Land* (Philadelphia: L. Johnson, 1833), 9.

32. Reproduced in Mary Anne Hines et al., *The Larder Invaded: Reflections on Three Centuries of Philadelphia Food and Drink* (Philadelphia: Library Company of Philadelphia and Historical Society of Pennsylvania, 1987), 25.

33. "Six Hundred Busy Candy-Making Maidens," *Soda Fountain,* March 1906, 22.

34. Coe and Coe, *True History,* 170.

35. Henry Weatherley, *A Treatise on Boiling Sugar, Crystallizing, Lozenge-Making, Comfits, Gum Goods, and other Processes for Confectionery, Etc.* (Philadelphia: H. C. Baird, 1865), 607.

36. The Fishers were a well-known aristocratic Philadelphia family who enjoyed the finer things in life. Mrs. Fisher was most likely married to either Sidney George or Joshua Francis Fisher, both of whom were independently wealthy in addition to being plantation owners.

37. Stanley Lebergott, *Manpower and Economic Growth: The American Record since 1800* (New York: McGraw-Hill, 1964), 541.

38. "Christmas," *Charleston Mercury,* 23 December 1861.

39. "Cakes and Confectionery," *Delaware County Republican,* 15 December 1865.

40. "A Novelty in Philadelphia Manufactures," *Delaware County Republican,* 26 December 1866.

41. J. W. Parkinson, Receipt, 8 February 1853, Library Company of Philadelphia; L. N. Smith, Receipt, 1869, Henry Francis Du Pont Winterthur Museum and Library; N. S. Dickey & Co., *Special Price-List,* 1 January 1879, Thomas Mills Collection, Mercer Museum, Doylestown, Pennsylvania.

42. Croft & Wilbur Co., *List of Prices of Confections,* 1876, Library Company of Philadelphia; John M. Miller & Son, Price List, 1876, Thomas Mills Collection, Mercer Museum, Doylestown, Pennsylvania; N. S. Dickey & Co.; *Special Catalogue, Felix Potin, Paris* (New York: Strauss & Lee, 1889).

43. *Frank Leslie's Historical Register of the United States Centennial Exposition, 1876,* ed. Frank Henry Norton (New York: Frank Leslie's Publishing House, 1877), 348.

44. Ibid., 212.

45. Julia Davis Chandler, "Some Peppermint Dainties," *Boston Cooking School Magazine* 7 (June/July 1902–May 1903): 146.

46. *The Candy-Maker* (New York: Excelsior Publishing House, 1896), 478.

47. Walter M. Lowney Co., *Report of the Lowney Educational Convention* (n.p.: n.p., 1911), 14.

48. "Six Hundred Busy Candy-Making Maidens," 22.

49. "Costly Bonbon Boxes," *Confectioners' and Bakers' Gazette,* 10 March 1902, 22.

50. Letter from Henry Maillard to William Arthur Brieseu, 15 July 1906. Warshaw Collection of Business Americana. Archives Center, National Museum of American History, Smithsonian Institution.

51. Edward Young, *Labor in America: Showing the Rates of Wages and Cost of Subsistence in the United States and British America, in the Year 1874, as Compared with Previous Years* (Washington, D.C.: Government Printing Office, 1875), 62. The statistic is for superfine wheat flour in New York for 1874. P. Arnaud, Receipt 22 May 1877, Henry Francis Du Pont Winterthur Museum and Library.

52. Tiffany & Co. Archives, Parsippany, New Jersey, item # A.1999.54.

53. "A Varied Assortment," *Confectioner and Baker,* 15 December 1901, 8.

54. Historicus [Richard Cadbury], *Cocoa: All About It* (London: Sampson Low, Marston, 1896), 65.

55. "Costly Bonbon Boxes," 22.

56. Terrio, *Crafting,* 257.

57. Commodities are objects having concrete "exchange value," material equivalencies at the moment of exchange. Gifts, on the other hand, pull one into a cycle of obligations. For more on this relationship, see Thomas, *Entangled Objects;* Daniel Miller, *Material Culture and Mass Consumption* (New York: Blackwell, 1994); Celia Lury, *Consumer Culture* (New Brunswick, N.J.: Rutgers University Press, 1996); and Karl Marx, *Capital: A Critique of Political Economy,* vol. 1 (1867; New York: Vintage Books, 1977).

58. George H. Hazlitt, *Historical Sketch of the Confectionery Trade of Chicago* (Chicago: Jobbing Confectioners' Association, 1905), 142.

59. Mrs. Eliza Duffey, *The Ladies' and Gentlemen's Etiquette* (Philadelphia: Porter & Coates, 1877), 150.

60. Mrs. Oliver Bell Bunce, *What to Do* (New York: D. Appleton, 1892), 60.

61. Francis J. Grund, "Christmas and New Year in France and Germany," *Godey's Lady's Book* 36 (January 1848): 8.

62. Diane Barthel, "Modernism and Marketing: The Chocolate Box Revisited," *Theory, Culture and Society* 6 (1989): 434.

63. James H. Barnett, "The Easter Festival—A Study in Cultural Change," *American Sociological Review* 14 (1949): 68–69. Perhaps overstating her case, Barthel writes, "Chocolate is part of what Bataille recognized as that part of life that is *excessive:* extra, surplus, having more to do with losing control than with gaining it, with spending rather than saving, with sex rather than salvation" ("Modernism and Marketing," 437).

64. Charles Beezley, *Our Manners and Social Customs: A Practical Guide to Deportment, Ease, Manners, and Social Etiquette* (Chicago: Elliott & Beezley, 1891), 82.

65. "Advertising the Little Candy Store," *Confectioners' and Bakers' Gazette,* 10 August 1905, 25.

66. Nicholas Thomas has written that "the artifact is not simply a valuable object of exchange or even a gift that creates relations of one sort or another, but also a crucial index of the extent to which those relations are sustained or disfigured" (*Entangled Objects,* 19).

67. "Sweets as a Factor in Civilization," *Confectioner and Baker,* January 1902, 19.

68. Ibid., 18.

69. Textural qualities often separated good chocolates and bonbons from bad. Those made with additives or adulterants not only lacked taste but also did not melt as smoothly in the mouth: "Many readers who have bought the common fruit or nut-candy, will remember that in letting the round pieces containing the fruit melt in their mouths, the result was a thick tasteless paste without taste or flavor" (*Candy Making at Home, by One Who Has Tried It* [Denver: Perry Publishing Co., 1887], 10–11).

70. Hazlitt, *Historical Sketch,* 142.

71. "Love Candy," *Confectioners' and Bakers' Gazette,* 10 March 1906, 28.

72. "Impressions Windows Create," *Confectioners' and Bakers' Gazette,* 10 June 1902, 17.

73. Leigh Eric Schmidt, "The Fashioning of a Modern Holiday: St. Valentine's Day, 1840–1870," *Winterthur Portfolio* 28, 4 (1993): 233.

74. Leigh Eric Schmidt, *Consumer Rites: The Buying and Selling of American Holidays* (Princeton: Princeton University Press, 1995), 96.

75. Quoted in ibid., 95–96.

76. "A Neat Advertisement," *Confectioner and Baker,* March 1903, 6.

77. "Sweets as a Factor in Civilization," 18.

78. Historicus, *Cocoa,* 43.

79. Coe and Coe, *True History,* 99.

80. T. J. Jackson Lears, *Fables of Abundance: A Cultural History of Advertising in America* (New York: Basic Books, 1994), 146, 148. See also, Michael Sturma, "The Nubile Savage," *History Today* 45 (April 1995): 7–9; and Anne McClintock, *Imperial Leather: Race, Gender, and Sexuality in the Colonial Conquest* (New York: Routledge, 1995).

81. Bernard Minifie, *Chocolate, Cocoa and Confectionery: Science and Technology* (New York: Van Nostrand Reinhold, 1989), 5.

82. Coe and Coe, *True History,* 95.

83. Henry Stubbes, quoted in ibid., 176–77.

84. Giovanni Bianchi, quoted in ibid., 208.

85. Quoted in Schivelbusch, *Tastes of Paradise,* 94.

86. James Wadsworth, *A Curious History of the Nature and Quality of Chocolate,* quoted in Linda K. Fuller, *Chocolate Fads, Folklore, and Fantasies* (New York: Haworth Press, 1994), 137.

87. Eliza Haywood, *A Present for a Servant-Maid* (London: T. Gardner, 1743), 21.

88. Claude Fischler, "Attitudes Towards Sugar and Sweetness in Historical and Social Perspective," in *Sweetness,* ed. John Dobbing (London: Springer-Verlag, 1987), 90.

89. Margaret Coxe, *The Young Lady's Companion: In a Series of Letters* (Columbus, Ohio: I. N. Whiting, 1839), 217–18, 321–22, 318.

90. Karen Lystra, *Searching the Heart: Men, Women, and Romantic Love in Nineteenth-Century America* (New York: Oxford University Press, 1989), 106.

91. J. H. Kellogg, *Plain Facts for Old and Young: Embracing the Natural History and Hygiene of Organic Life* (1888; reprint, Arno Press, 1974), 232, 237, 245.

92. Ibid., 408. Other authors also linked eating bonbons and other vices. One account reported, "A lady recently went into a confectionery store to purchase some bonbons. She was handsomely dressed, and was quite pretty. As the proprietor was making up her parcel he saw her stagger and fall. Hastening round to the front of the counter, he found her lying helpless on the floor, dead drunk" (James D. McCabe, *Secrets of the Great City* [Philadelphia, 1868], 371).

93. George Hillard, *The Dangers and Duties of the Mercantile Profession: An Address Delivered Before the Mercantile Library Association, at Its Thirtieth Anniversary, November 13, 1850* (Boston: Ticknor & Fields, 1854), 23–24. I would like to thank Joe Rainer for this citation.

94. Historicus, *Cocoa,* 30–31.

95. William A. Alcott, *Familiar Letters to Young Men on Various Subjects* (Buffalo: G. H. Derby, 1850), 73–74.

96. William Greenleaf Eliot, *Lectures to Young Men* (Boston: Unitarian Association, 1867), 74.

97. Joseph P. Tuttle, *The Way Lost and Found* (Philadelphia: Presbyterian Board of Publication, 1870), 225–26.

98. George Horner, *A Manual on Beauty and Health* (Providence, R.I.: B. T. Albro, 1848), 20–21.

99. Coxe, *Young Lady's Companion,* 217–18.

100. Robert Tomes, *The Bazar Book of Decorum* (New York: Harper & Brothers, 1871), 26–27.

101. Censor, *Don't, or, Directions for Avoiding Improprieties in Conduct and Common Errors of Speech* (New York: D. Appleton, 1884), 90.

102. Tomes, *Bazar,* 193–94.

103. *Boston Cooking School Magazine* 1, 2 (1896): 88–89; emphasis in original.

104. Quoted in William Randel, "John Lewis Reports the Centennial," *Pennsylvania Magazine of History and Biography* 79 (1955): 365; emphasis in original.

105. Werner Sombart, *Luxury and Capitalism* (1913; Ann Arbor: University of Michigan Press, 1967), 60–61.

106. *Cocoa and Chocolate: A Short History of Their Production and Use* (Dorchester, Mass.: Walter Baker, 1904), no pagination; emphases in original.

107. Alfred J. H. Crespi, "Cocoa and Chocolate," *Gentleman's Magazine* 269 (October 1890): 374.

108. *The Chocolate-Plant (Theobroma cacao) and Its Products* (Dorchester, Mass.: Walter Baker, 1891), 35.

109. See Joseph Richard Snavely, *Milton S. Hershey, Builder* (Hershey, Pa.: J. R. Snavely, 1935).

110. "Another Best-Eating Milk Chocolate," *International Confectioner,* November 1905, 26.

111. Timothy M. Erdman, "Hershey: Sweet Smell of Success," *American History Illustrated* 29 (March–April 1994): 68.

112. "Wonderful Chocolates," *Confectioner and Baker,* March 1901, 19.

113. Snavely, *Milton S. Hershey,* 6; Roy Bogartz, "The Chocolate Camelot," *American Heritage* 24, 4 (June 1973): 7. A colleague informs me that he has seen the phrase "French Secrets" used in ante-bellum birth control advertisements as well.

114. Schmidt, *Consumer Rites,* 219, 223.

115. "Easter Novelties," *Confectioners' and Bakers' Gazette,* 10 April 1902, 19.

116. "Easter Specialties," *Confectioner and Baker,* May 1902, 22, 23.

117. "How Easter Eggs Are Made by the Thousand," *Scientific American,* 14 April 1906, 308.

118. Elizabeth Clark Kieffer, "Easter Customs of Lancaster County," *Papers of the Lancaster Historical Society* 52 (1948): 60.

119. Barnett, "Easter Festival," 68.

120. Schmidt, *Consumer Rites,* 224–25.

121. "Visit to a Chocolate Factory," *Chambers' Edinburgh Journal,* 2d ser., 2 (July–

December 1852): 282. For more about women in confectionery factories, see Helen L. Sumner, *Report on the Condition of Woman and Child Wage-Earners in the United States,* vol. 10: *History of Women in Industry in the United States* (Washington, D.C.: Government Printing Office, 1910), 189–91; Elizabeth Beardsley Butler, *Women and the Trades, Pittsburgh, 1907–1908* (1911; reprint, Arno Press, 1969), 44–59; Cornelia Stratton Parker, *Working with the Working Woman* (New York: Harper & Bros., 1922), 1–41; and Gail Cooper, "Love, War, and Chocolate: Gender and the American Candy Industry, 1890–1930," in *His and Hers: Gender, Consumption, and Technology,* ed. Roger Horowitz and Arwen Mohun (Charlottesville: University Press of Virginia, 1998), 67–94.

122. "Visit to a Chocolate Factory," 282.

123. A. W. Knapp, *Common Commodities and Industries: Cocoa and Chocolate* (London: Sir Isaac Pitman & Sons, ca. 1923), 130–32.

124. "Half an Hour in a Candy Factory," *Confectioners' and Bakers' Gazette,* 10 February 1902, 20.

125. James S. Wilson, *Modern Candy Making: How to Make All Kinds of Candy* (Chicago: G. W. Ogilvie, 1904), 7–8.

126. Hazlitt, *Historical Sketch,* 127.

127. Edith A. Browne, *Peeps at Industry: Cocoa* (London: A. & C. Black, 1920), 87.

128. J. M. Sanderson, *The Complete Confectioner, Pastry-Cook, and Baker* (Philadelphia: Lea & Blanchard, 1844), 85.

129. National Equipment Co., *Catalogue* (Springfield, Mass.: National Equipment Co., ca. 1920), 134, 136.

130. Ibid., 140.

FIVE ⁓ *The Icing on the Cake*

1. "To Be Seen," *Pennsylvania Gazette,* 20 June 1765.

2. Ibid.

3. Ice pyramids, cousins of sugar sculptures, began their popularity in seventeenth-century Italy. See Elizabeth David, *Harvest of the Cold Months: The Social History of Ice and Ices* (New York: Viking, 1994), esp. 61–62, 100–101.

4. Madeleine Pelner Cosman, *Fabulous Feasts: Medieval Cookery and Ceremony* (New York: George Braziller, 1976), 31; see also Terence Scully, "Mediaeval French *Entremets,*" *Petits Propos Culinaires* 17 (June 1984): 44–56.

5. Cosman, *Fabulous Feasts,* 33.

6. Bridget Ann Henisch, *Fast and Feast: Food in Medieval Society* (University Park: Pennsylvania State University Press, 1976), 231–32.

7. Quoted in "President Washington in New York, 1789," *Pennsylvania Magazine of History and Biography* 32 (1908): 499.

8. Louise Belden, *The Festive Tradition: Table Decoration and Desserts in America, 1650–1900* (New York: Norton, 1983), 79.

9. "Confectionary," *American Messenger,* 27 September 1837, 4.

10. Victoria de Grazia, "Introduction," in *The Sex of Things: Gender and Consumption*

in Historical Perspective, ed. id., with Ellen Furlough (Berkeley: University of California Press, 1996), 19.

11. James Page, *Guide for Drawing the Acanthus, and Every Description of Ornamental Foliage* (London: Atchley, 1840); J. W. Parkinson, *The Complete Confectioner, Pastry-Cook, and Baker* (Philadelphia: Lea & Blanchard, 1844), 85.

12. See John Kasson, *Civilizing the Machine: Technology and Republican Values in America, 1776–1900* (New York: Penguin Books, 1976).

13. *The Industries of Philadelphia* (Philadelphia: Richard Evans, 1881), 211.

14. "Great Railroad Banquet and Ball," *Frank Leslie's Illustrated Weekly,* 26 April 1856, 311.

15. "News from Washington," *New York Herald,* 6 February 1862.

16. "The Grand Presidential Party," *New York Herald,* 5 February 1862.

17. Ibid.

18. Stuart Bruce Kaufman, *A Vision of Unity: The History of the Bakers and Confectionery Workers International Union* (n.p.: University of Illinois Press, 1986), 7.

19. John Hounihan, *Bakers' and Confectioners' Guide and Treasure* (Staunton, Va.: John D. Hounihan, 1877), 34; Paul Brenner, "The Formative Years of the Hebrew Bakers' Union, 1881–1914," *YIVO Annual of Jewish Social Science* 15 (1983): 42; Marie-Antoine Carême, *The Royal Parisian Pastrycook and Confectioner* (London: Mason, 1834), "Pièces Montées."

20. J. Thompson Gill, *The Complete Practical Ornamenter* (Chicago: Caterer Publishing Co., 1891), preface.

21. T. J. Jackson Lears, *No Place of Grace: Antimodernism and the Transformation of American Culture, 1880–1920* (Chicago: University of Chicago Press, 1981), 147–48.

22. Gill, *Complete Practical Ornamenter,* preface.

23. For more on this, see Colin Campbell, *The Romantic Ethic and the Spirit of Modern Consumerism* (New York: Blackwell, 1987), esp. 92.

24. *A Sylvan City: Or, Quaint Corners in Philadelphia* (Philadelphia: Our Continent Publishing Co., 1883), 252.

25. "Preparing for the Holidays," *Frank Leslie's Illustrated Weekly,* 13 January 1872, 279.

26. C. H. King, "The Village Church," *Bakers Review,* 15 December 1905, 35.

27. This was consonant with acceptable ideas of clothing at the time as well. A garishly dressed woman was considered ill-bred and a bit of a trollop.

28. C. H. King, "Illustration of Rock Work," *Bakers Review,* 15 March 1903, 31.

29. Agnes Carr Sage, "The Evolution of the Wedding Cake," *Lippincott's Magazine* 57 (January–June 1896): 410.

30. Simon Charsley, *Wedding Cakes and Cultural History* (London: Routledge, 1992), 139.

31. "Pound Cake," *Lady's Book* 2 (May 1831): 271.

32. "Bridal Cake at Rensellaer Manor," *Illustrated News,* 19 February 1853, 125.

33. "Festal Cakes," *All the Year Round* 40 (12 February 1887): 79–82.

34. Francis J. Ziegler, "Ceremonial Cakes," *Cosmopolitan* 29 (May 1900): 33.

35. Emily Thornwell, *The Lady's Guide to Perfect Gentility* (Philadelphia: J. B. Lippincott, 1868), 104. For accounts of eighteenth-, nineteenth-, and early twentieth-century

weddings, see Allan Kulikoff, "Throwing the Stocking: A Gentry Marriage in Provincial Maryland," *Maryland Historical Quarterly* 71, 4 (Winter 1976): 516–21; Solomon L. Loewen, "Courting and Weddings among the Early Settlers in Ebenfield Community," *Journal of the American Historical Society of Germans from Russia* 7, 3 (1984): 21–23; Bruce Daniels, "Frolics for Fun: Dances, Weddings, Dinner Parties in Colonial New England," *Historical Journal of Massachusetts* 21, 2 (1993): 1–22; "A Wedding at Independence, California in 1876," *Western States Jewish History* 22, 2 (1990): 112–15; "The Elegant Wedding of a Pioneer's Daughter," ibid. 17, 2 (1985): 160–62; Seebert Goldowsky, "Society Wedding," *Rhode Island Jewish Historical Notes* 10, 3 (1989): 317–22; and Lionel Heynemann, "The Edward Brandenstein–Florine Haas Wedding, San Francisco, 1903," *Western States Jewish Historical Quarterly* 15, 2 (1983): 184–86.

36. "Marriage of Princess Louise at Windsor," *Illustrated London News,* 1 April 1871, 308.

37. "Monster Wedding Cakes," *Confectioner and Baker,* June 1901, 20.

38. Ibid.

39. Marcia Seligson, *The Eternal Bliss Marriage: America's Way of Wedding* (New York: William Morrow, 1973), 98.

40. Charsley, *Wedding Cakes,* 97.

41. Thorstein Veblen, *The Theory of the Leisure Class: An Economic Study of Institutions* (1899; New York: Mentor Books, 1953), 126.

42. "Elementary Icing," *Confectioner and Baker,* 15 February 1902, 1–2.

43. "The Bride's Loaf and Wedding Cake," *Boston Cooking School Magazine* 8 (June/July 1903–May 1904): 42.

44. *Selections for Autograph and Writing Albums* (New York: Chas. A. Lilley, 1879), 22.

45. Emily Holt, *Encyclopedia of Etiquette* (New York: Syndicate Publishing Co., 1915), 203.

46. *Weddings and Wedding Anniversaries* (New York: Butterick, 1894), 15.

47. Gill, *Complete Practical Ornamenter,* 32.

48. There is one such piece located at the Albany Institute of History and Art, belonging to Albertine van Alstyne, which came from her marriage to Frederick Fitch in 1834. Another, supposedly a piece from Tom Thumb's marriage in 1863, was recently auctioned off in North Carolina. I would like to thank Sandra Markham, of the Albany Institute of History and Art, for bringing these examples to my attention.

49. *Weddings and Wedding Anniversaries,* 14.

50. Charsley, *Wedding Cakes,* 126.

51. Ziegler, "Ceremonial Cakes," 33.

52. Mary Manners, "The Unemployed Rich," *Everybody's Magazine,* 9 (1903): 44.

53. "Costly Wedding Cakes: Art in Confectionery," *Confectioners' Journal,* June 1890, 66.

54. "Wedding Cake and Slices," *Bakers Review,* 15 September 1904, 65.

55. For more discussion about this, see Gwendolyn Wright, *Moralism and the Model Home: Domestic Architecture and Cultural Conflict in Chicago, 1873–1913* (Chicago: University of Chicago Press, 1980), esp. 81–97.

56. Gill, *Complete Practical Ornamenter,* preface.

57. Ian McKay, "Capital and Labour in the Halifax Baking and Confectionery Industry During the Last Half of the Nineteenth Century," *Labour* 3 (1978): 65.

58. T. Percy Lewis and A. G. Bromley, *The Victorian Book of Cakes* (1902; New York: Portland House, 1991), 53.

59. Charsley, *Wedding Cakes,* 94–95.

60. Gill, *Complete Practical Ornamenter,* preface.

61. Ibid., 30–31.

62. Lewis and Bromley, *Victorian Book of Cakes,* 50–51.

63. C. H. King, "Ornamental Work for Beginners," *Bakers Review,* 15 January 1904, 41.

64. Felix Adler quoted in Viviana Zelizer, *Pricing the Priceless Child: The Changing Social Value of Children* (New York: Basic Books, 1985), 70.

65. Lewis and Bromley, *Victorian Book of Cakes,* 40–41.

SIX ⁘ *Home Sweet Home*

1. *One Hundred Years' Progress of the United States* (Hartford, Conn.: L. Stebbins, 1871), 129.

2. Mary Randolph, *The Virginia Housewife* (Baltimore: Plaskitt & Cugle, 1839), xii.

3. A search in the RLIN (Research Libraries Group) database shows that the publication of English-language confectionery manuals peaked at the turn of the century. Decade by decade numbers are as follows: 1800–1809: 6; 1810–19: 5; 1820–29: 7; 1830–39: 12; 1840–49: 7; 1850–59: 1; 1860–69: 9; 1870–79: 13; 1880–89: 29; 1890–99: 35; 1900–1909: 93; 1910–15: 42. While the RLIN database is not exhaustive, it does reveal broad publishing trends. These figures do not include general cookbooks, which also included recipes for making candies and desserts; nor do they include advertising recipe pamphlets, which were published by the hundreds (if not thousands) from the 1880s to World War I. My personal collection of confectionery-related advertising recipe pamphlets from this era alone numbers over fifty. See also Katherine Golden Bitting, *Gastronomic Bibliography* (San Francisco: n.p., 1939) and Eleanor Lowenstein, *Bibliography of American Cookery Books, 1742–1860* (Worcester, Mass.: American Antiquarian Society, 1972).

4. Helen Campbell, *The American Girl's Home Book of Work and Play* (New York: G. P. Putnam's Sons, 1883), 407.

5. T. J. Jackson Lears, *No Place of Grace: Antimodernism and the Transformation of American Culture, 1880–1920* (Chicago: University of Chicago Press, 1981), 16.

6. Campbell, *American Girl's Home Book,* 407.

7. See, e.g., Mary Hinman Abel, *Sugar as Food,* Farmer's Bulletin No. 93, U.S. Department of Agriculture (Washington, D.C.: Government Printing Office, 1899), 10.

8. Sarah Tyson Rorer, "Dietetic Sins and Their Penalties," *Ladies' Home Journal,* January 1906, 42.

9. Christine Herrick, *Candy Making in the Home* (Chicago: Rand McNally, 1914), 5–6.

10. Maude C. Cooke, *Twentieth Century Cook Book* (n.p.: n.p., [189?]), 541.

11. *Candy Making at Home, by One Who Has Tried It* (Denver: Perry Publishing Co., 1887), 9–10.

12. *Dr. Miles' Candy Book* (Elkhart, Ind.: Miles Medical Co., ca. 1890), 3.

13. *Candy Making at Home,* 9–10.

14. Janet M. Hill, "After Breakfast Chat," *Boston Cooking School Magazine* 8 (June/July 1903–May 1904).

15. "After-Breakfast Chats with Young Housekeepers," *Boston Cooking School Magazine* 2 (June/July 1897–April/May 1898): 223.

16. *One Hundred Years' Progress of the United States,* 129; Harvey Wiley, *Foods and Their Adulteration: Origin, Manufacture, and Composition of Food Products; Infants' and Invalids' Foods; Detection of Common Adulterations, and Food Standards* (1907; 2d ed. Philadelphia: P. Blakiston's Son, 1911), 455.

17. Population statistics based on 1800 and 1900 census figures, which figure population at 5,084,912 and 74,607,225, respectively (U.S. Historical Census Database; Intra-University Consortium for Political and Social Research).

18. Mary Elizabeth Hall, *Candy-Making Revolutionized; Confectionery from Vegetables* (New York: Sturgis & Walton, 1912), ii; emphasis added. The author spent a great deal of her book trying to convince women how delectable her concoctions were. Of "Lima Bean Taffy," she wrote: "Bean taffy easily takes first rank among all taffies—vegetable or otherwise. The taste is good beyond words, and the consistency is pleasingly 'chewy' without being tenacious to the point of teeth pulling!" (103).

19. G. V. Frye, *The Housewife's Practical Candy Maker* (Chicago: Belford, Clarke, 1889), 3–4.

20. *Our Reliable Candy Teacher Has Entirely New Ideas in the Art of Home Candy Making* (Conshocton, Ohio: Hissong Candy School, ca. 1911), iii.

21. Nelle M. Mustain, *Popular Amusements for In and Out of Doors* (Chicago [?]: L. A. Martin, ca. 1902), 331.

22. Thorstein Veblen, *The Theory of the Leisure Class* (1899; New York: Mentor Books, 1953), 69.

23. M. E. Rattray, *Sweetmeat-Making at Home* (London: C. Arthur Pearson, 1922), x–xiii.

24. M. E. W. S., *Home Amusements* (New York: D. Appleton, 1881), 142.

25. *They Wanted Jell-O* (Leroy, N.Y.: Genesee Pure Food Co., ca. 1907).

26. Virginia Penny, *How Women Can Make Money, Married or Single* (Philadelphia: John E. Potter [1862]), 155.

27. Statistics taken from Gopsill's *Philadelphia City Directory* (Philadelphia: J. Gopsill's Sons, 1850); *Boyd's Co-Partnership and Residential Business Directory of Philadelphia City* (Philadelphia: T. K. Collins, Jr., 1859–60, 1874, 1881).

28. See, e.g., Campbell, *American Girl's Home Book,* 407; Ellye Howell Glover, *"Dame Curtsey's" Book of Candy Making* (Chicago: A. C. McClurg, 1913), foreword; and Herrick, *Candy Making,* 14–15. *The Art of Confectionery* (Boston: J. E. Tilton, 1866), suggested that while "all ladies who possess a taste and genius for this beautiful art" could make confec-

tionery that would "not be disdained by the highest minds" (13), a "lucrative and pleasant business might thus be opened to farmers' wives and daughters especially, who could easily prepare, from the abundance that surrounds them, fine confectionery for market where it would meet a ready sale" (14). Even though rich women did not sell their candy, people thought it logical for someone like the farmer's wife or daughter, already a laborer, who could easily supplement the family income that way.

29. Campbell, *American Girl's Home Book,* 407.

30. *The Art of Home Candy Making* (Canton, Ohio: Home Candy Makers, 1915).

31. Even though the percentage of female confectioners was very high, especially in the 1880s and 1890s, most of these women failed to establish long-standing businesses. City directories indicate a high turnover rate: female confectioners appear year after year, but very few held the occupation for consecutive years.

32. Sarah Tyson Rorer, *Home Candy Making* (Philadelphia: Arnold, 1889), 6.

33. See, e.g., Anne Cooper Funderburg, *Chocolate, Strawberry, and Vanilla: A History of American Ice Cream* (Bowling Green, Ohio: Bowling Green State University Popular Press, 1995), 40–49.

34. Oscar Edward Anderson, *Refrigeration in America: A History of a New Technology and Its Impact* (1953; Port Washington, N.Y.: Kennikat Press, 1972), 54.

35. *The Housewife's Library,* ed. George A. Peltz (n.p.: Englewood Publishing, 1885), 295.

36. Ice "caves" were, in fact, used fairly successfully to store ice cream for up to two hours. Different from ice boxes, which were proto-refrigerators with the block of ice kept in a separate compartment from the food, ice caves were small insulated boxes. Filled with ice, which surrounded the molded ice cream, they kept the ice cream fairly solid while it was "curing" or "ripening" before serving.

37. *Housewife's Library,* ed. Peltz, 293–94.

38. "A Sermon on Ice Cream," *Caterer* 1 (1882–83): 423.

39. Ice cream freezer liners were usually made of copper coated with tin. When the tin wore off, the copper was directly exposed to the ice cream, causing what was called "verdigris poisoning." For a contemporary account of alleged verdigris poisoning among customers of a reputable confectioner, see "Verdigris Ice Cream," *Caterer* 2 (1883–84): 420–22.

40. Cornelius Weygandt, *Philadelphia Folks: Ways and Institutions in and about the Quaker City* (New York: D. Appleton, 1938), 19.

41. David W. Miller, "Technology and the Ideal: Production Quality and Kitchen Reform in Nineteenth-Century America," in *Dining in America, 1850–1900,* ed. Kathryn Grover (Amherst: University of Massachusetts Press; Rochester, N.Y.: Margaret Woodbury Strong Museum, 1987), 66–68.

42. Weygandt, *Philadelphia Folks,* 20.

43. "Ices," *Boston Cooking School Magazine* 1 (Summer/Autumn 1896–Winter/Spring 1897): 22.

44. Bernard Lyman, *A Psychology of Food: More Than a Matter of Taste* (New York: Van Nostrand Reinhold, 1989), 104.

45. "Query No. 56," *Boston Cooking School Magazine* 2 (June/July 1897–April/May 1898): 179.

46. Charles Marine Metzgar, *Home-Made Ice Cream and Candy* (Chicago: Metzgar Publishing Co., 1908), 3.

47. *Jell-O Ice Cream Powder* (Leroy, N.Y.: Genesee Pure Food Co., 1915); emphases in original.

48. *Makes Ice Cream: Jell-O Ice Cream Powder* (Leroy, N.Y.: Genesee Pure Food Co., 1905).

49. Philip P. Gott, *All About Candy and Chocolate: A Comprehensive Study of the Candy and Chocolate Industries* (Chicago: National Confectioners' Association of the United States, 1958), 149.

50. *How to Make Candy* (Hartford, Conn.: N. P. Fletcher, 1875), 5–6.

51. Kate Gannett Wells, "Little Gifts Make Great Friendships," *Boston Cooking School Magazine* 9 (June/July 1904–May 1905): 438.

52. Julia Davis Chandler, "Fudge Up to Date," *Boston Cooking School Magazine* 3 (June/July 1898–April/May 1899): 274.

53. Glover, *"Dame Curtsey's,"* 11.

54. Ralph Richmond, *Samplers: Their Story . . . as Told Through the Whitman's Collection* (Philadelphia: Whitman's Chocolate Division, Pet Inc., 1970), 17.

55. For more on this idea, see Diane Barthel, "Modernism and Marketing: The Chocolate Box Revisited," *Theory, Culture and Society* 6 (1989): 434.

56. Richmond, *Samplers,* 17; emphasis in original.

57. Kenneth Ames, *Death in the Dining Room and Other Tales of Victorian Culture* (Philadelphia: Temple University Press, 1992), 127.

58. Ames, *Death in the Dining Room,* 115.

59. Richmond, *Samplers,* 17.

60. *Art of Home Candy Making,* 58.

61. Rattray, *Sweetmeat-Making,* 146.

62. N. C. Hall, *Directions for Making Chocolate Candy at Home with Hall's Home Outfit* (Lyme, Conn.: N. C. Hall, 1915), 17.

63. *How to Make Candy,* 69.

64. Annie M. Jones, *Homespun Candies: Simple and Thoroughly Tested Recipes for Candy to be Made at Home* (New York: Rohde & Haskins, 1904), 8.

65. Abel, *Sugar as Food,* 17.

66. Crandall Shifflett, *Victorian America, 1876–1913* (New York: Facts on File, 1996), 63.

67. Sidney W. Mintz, *Sweetness and Power: The Place of Sugar in Modern History* (New York: Penguin Books, 1985), 194.

68. *Dr. Miles' Candy Book,* 3.

69. Glover, *"Dame Curtsey's,"* 57.

70. My own study of eighteenth- and nineteenth-century manuscript recipe books (hand-written culinary journals kept by women before the advent of recipe cards) has revealed a high percentage of sweet and baked/farinaceous recipes. Sweet dishes alone made up more than half of the recorded recipes in eight different Philadelphia-area

books. In addition, many of these books contained recipes transcribed from others, which were then "rated"—commented on in the accompanying marginalia. This study found a consistent presence of and importance placed on sweet dishes, and it also showed that the women evaluated their peers based on the success of these sweet dishes.

71. *The Art of Confectionery* (Boston: J. E. Tilton, 1866), 13.

72. "Query No. 74," *Boston Cooking School Magazine* 2 (June/July 1897–April/May 1898): 304.

73. "Query No. 76," *Boston Cooking School Magazine* 2 (June/July 1897–April/May 1898): 305; emphasis in original.

74. The breakdown of recipe types in the dessert category from the *Ladies' Home Journal* by percentage is as follows: cakes, 21.31; puddings, 19.93; fruits, 13.22; candies, 11.84; frozen desserts, 6.46; cookies, 6.11; jellies and jams, 6.06; pastries, 5.57; biscuits, 5.18; icings and cake fillings, 2.56; and fritters, 1.33. These figures and those for the manuscript recipe books are results of my own personal survey.

75. "Sweets Good for the Teeth," *Boston Cooking School Magazine* 2 (June/July 1897–April/May 1898).

76. "Desserts," *Boston Cooking School Magazine* 1 (Summer/Autumn 1896–Winter/Spring 1897): 199.

77. M. E. Converse, "Three 'Goodies,'" *Boston Cooking School Magazine* 3 (June/July 1898–April/May 1899): 289.

78. What manuscript recipe books there are from the turn of the century show an increasing inclusion of brand names in transcribed recipes. In addition, women integrated more blatant advertising material, pasting printed recipes that incorporated commercial products into the leaves of their books.

79. A typical traditional recipe for Calf's Foot Jelly reads:

> Boil two feet in two quarts and a pint of water till the feet are broken and the water half wasted; strain it, and when cold, take off the fat, and remove the jelly from the sediment; then put it into a saucepan, with sugar, raisin wine, lemon-juice to your taste, and some lemon-peel. When the flavour is rich, put to it the whites of two eggs well beaten, and their shells broken. Set the saucepan on the fire, but do not stir the jelly after it begins to warm. Let it boil twenty minutes after it rises to a head; then pour it through a flannel jelly-bag first dipping the bag in hot water to prevent waste, and squeezing it quite dry. Run the jelly through and through until clear; then put it into glasses or forms. (Maria Eliza Rundell [Ketelby], *American Domestic Cookery.* [Baltimore: Matchett for Lucas, 1819])

80. I know this from replicating Maria Rundell's recipe myself.

81. *Jell-O: America's Best Family Dessert* (Leroy, N.Y.: Genesee Pure Food Co., 1902).

82. *Jell-O Ice Cream Powder* (Leroy, N.Y.: Genesee Pure Food Co., 1906), 1; *Jell-O the Dainty Dessert* (Leroy, N.Y.: Genesee Pure Food Co., 1905), 1.

83. *Jell-O* (Leroy, N.Y.: Genesee Pure Food Co., 1904).

84. Rose Markward, *Dainty Desserts for Dainty People* (n.p.: Charles B. Knox, 1896), 1.

85. *Dainty Dishes Made of Knox Gelatine* (Johnstown, N.Y.: Charles B. Knox, 1910), 3.

86. Janet M. Hill and Emma H. Crane, *Dainty Junkets* (Little Falls, N.Y.: Johan D. Frederiksen, 1915), 2.

87. *Jell-O* (Leroy, N.Y.: Genesee Pure Food Co., 1904), 1.

88. For a counterpoint to this, see William I. Miller, *The Anatomy of Disgust* (Cambridge, Mass.: Harvard University Press, 1997).

Postscript

1. Chris Lecos, "Our Insatiable Sweet Tooth," *FDA Consumer* 19 (October 1985): 25–26; Beatrice Trum Hunter, "Confusing Consumers about Sugar Intake," *Consumers' Research Magazine* 78, 1 (January 1995): 14–18; "Nutrition Notes," *Health,* September 1996, 32; Marian Uhlman, "A Sugar Debate That's Not So Sweet," *Philadelphia Inquirer,* 10 April 2000, A1, 7.

2. Lecos, "Our Insatiable Sweet Tooth."

3. Ibid.

4. Uhlman, "Sugar Debate," A7.

5. Cindy Maynard, "Getting the Scoop on Sugar," *Current Health* 24, 6 (2 February 1998): 21–25; Uhlman "Sugar Debate," A7.

6. Maynard, "Getting the Scoop."

7. Suzanne Domel Baxter, "Are Elementary Schools Teaching Children to Prefer Candies but Not Vegetables?" *Journal of School Health* 68, 3 (March 1998): 111–14. As a result of candy being used as a reward, the author conjectures, children are less likely to want fruits and vegetables. This familiarity with candy increases children's "neophobia," their fear of new foods, especially when used coercively: "using food contingencies to coerce elementary school children to eat their vegetables (e.g., 'If you eat your broccoli, then you can eat dessert.') may be teaching them to dislike vegetables and like dessert."

8. Dianna Marder, "Remember Turkish Taffy?" *Philadelphia Inquirer,* 24 February 1998, F5.

9. Rick Nichols, "Beyond Godiva: Chocolate Memories," *Philadelphia Inquirer,* 28 June 2000.

10. Zlati Meyer, "Artisan Chocolatiers Capitalize on the Hand-made Look," *Philadelphia Inquirer,* 14 February 2001, C7.

11. *Spin,* June 2000.

12. Amanda Barnett, "Let Them Eat Cake: Imagination, Color Spice Up Traditional Wedding Dessert," CNN Interactive Writer, posted on 9 May 2000 at 11:28 A.M.

13. This chocolate cake was still the traditional shape—it had three tiers, and it was decorated with flowers and ivy. The woman who made the cake suggested these external embellishments to the bride, warning that the dark frosting alone might make the cake "look like a big pile of poo."

14. Barnett, "Let Them Eat Cake."

15. Claude Fischler, "Attitudes Towards Sugar and Sweetness in Historical and Social Perspective," in *Sweetness,* ed. John Dobbing (London: Springer-Verlag, 1987), 89.

16. *Oxford English Dictionary,* 2d ed. (Oxford: Clarendon Press, 1989), s.v.

17. Sidney Mintz began this project in "Sweet, Salt, and the Language of Love," *MLN* 106 (1991): 852–60.

18. Formal definitions and usages for these terms appear in one or more of the following sources: *Dictionary of American Slang,* ed. Robert L. Chapman (New York: HarperCollins, 1997); Paul Dickson, *Slang: The Authoritative Topic-by-Topic Dictionary of American Lingoes from All Walks of Life* (New York: Pocket Books, 1998); *The Dictionary of Contemporary Slang,* comp. Jonathan Green (New York: Stein & Day, 1985); and *The Random House Historical Dictionary of American Slang,* ed. J. E. Lighter (New York: Random House, 1994).

19. Dickson, *Slang.*

20. Mintz, "Sweet, Salt," 857–58.

21. Fischler, "Attitudes Towards Sugar and Sweetness," 89.

22. *Random House Historical Dictionary,* ed. Lighter, 710.

23. Alex Kuczynski, "'Arm Candy': One-Night Stun," *New York Times,* 27 September 1998, sec. 9, 1, 4.

24. I would like to thank David Miller for providing these examples and helping me understand the nuances of the meanings of sugar and sweetness in contemporary sports. See also Mintz's "Sweet, Salt" (858) for these and other examples of sweetness in speech.

ESSAY ON SOURCES

The following sources and repositories have shaped this book, reflecting the chorus of voices that have contributed to the evolving discussion of sugar and sweetness.

PRIMARY SOURCES

Just how and to what extent people used sugar can be teased from popular literature, with certain caveats. Cookbooks, medical treatises, etiquette manuals, magazine articles, and newspaper reports recorded everything from candy recipes to pictures of confections to warnings against overconsumption. Mostly prescriptive literature, such writings should always be considered with a critical eye, for they more often than not represent the interests of the authors. Yet the aggregation of consistent accounts reveals larger and broader trends among the American consuming public. The Rare Book and Special Collections Division at the Library of Congress maintains a large collection of early American cookbooks. The Winterthur Museum and Library not only has strong holdings of cookbooks, etiquette manuals, and other prescriptive literature but also contains a full run of the *Boston Cooking School Magazine,* with manuscript annotations by Janet McKenzie Hill, its original editor, showing the evolution of articles that appeared in print more than once. Other women's magazines, such as *Godey's Lady's Book* and, later, the *Ladies' Home Journal,* to be found at Winterthur and elsewhere, reflected as much as shaped contemporary fashions in domestic life. Periodicals such as these very often marshaled experts from the scientific community to write articles or serve as entrusted product spokespeople. The messages to women about popular attitudes toward sugar, especially in feeding it to their children, come through loud and clear in these publications. That Winterthur's collections extend into the twentieth century enabled me to trace the story of confectionery down to World War I. Winterthur's holdings of trade catalogues, which record the sequence of developing confectionery technologies and their viability within the changing marketplace, are also an invaluable source. The Hagley Museum in Wilmington, Delaware, and the Mercer Museum in Doylestown, Pennsylvania, each house business ledgers of Thomas Mills and Company, a particularly successful Philadelphia confectionery equipment manufacturer. These ledgers, along with other business materials, shed light on the commercial networks of confectionery equipment and, therefore, the production of certain types of confections all over the United States.

The Library Company of Philadelphia, which specializes in American print culture, houses all manner of relevant works representing the many voices I have quoted in this book. The breadth of its nineteenth-century literature in particular allowed me to look at sugar from a variety of perspectives. Popular ideas about gender propriety were articulated in books such as William Alcott's *Familiar Letters to Young Men* (Buffalo, ca. 1850), William Greenleaf Eliot's *Lectures to Young Men* (Boston, 1867), and Bronson Alcott's *The Physiology of Marriage* (Boston, 1856). The temperance angle was voiced in serials such as the *Temperance Advocate and Cold Water Magazine* (1840s), various temperance almanacs, and other similar publications.

The Library Company also contains countless published recipe books, household manuals, and domestic magazines, which in some cases were written by women. These books instructed women on how to run their homes and how to prepare economical, nutritious, or elegant meals for their guests and families. Important authors include Sarah Tyson Rorer, Elizabeth Rundell, and Hannah Glasse, all well-respected cookbook authors whose works appeared in many editions. Other publications contextualized the consumption of confectionery. City mysteries accomplished this most vibrantly, and often describe ice cream saloons and candy shops in deliciously sordid detail, painting them as sites of temptation and corruption. Quintessential examples of city mysteries include Solon Robinson's *Hot Corn* (New York, 1854), Ned Buntline's "The Death-Mystery: A Crimson Tale of Life in New York" (serialized in the *New York Mercury* in 1861), and George Lippard's *The Quaker City* (Philadelphia, 1844). Fundamentally humorous to us today, they reveal the general nineteenth-century anxiety about cities and urban anonymity. That confectionery represented yet one more potential danger to be found in America's burgeoning cities indicated that not everyone saw its consumption as a positive thing.

Literature about the history of African Americans supplied the passionate voices of abolitionists arguing against sugar consumption and the slave trade. In addition, this material expressed the view of self-interested merchants and politicians who argued for the institution's continuation. Even by the time Americans' sugar consumption had become a foregone conclusion, key reformers continued to express their concerns and discontent, believing that sugar was both a dietary and a moral scourge. Certain of these voices surfaced again and again throughout the century, providing a general barometer for both the growing prevalence of and opposition to sugar.

Trade journals, which most clearly reveal the various advertising strategies and market anxieties of professionals tied to the business of sweetness, also proved crucial to this study. The most varied and comprehensive holdings are located at the Library of Congress, where on good days staff people are willing to scan the stacks looking for errant volumes not represented on the catalogue cards. The variety of serials published indicates that sugar's ready availability supported all manner of professional endeavors from fancy cake decorating to penny candy manufacturing. Representative titles include the *Confectioners' and Bakers' Gazette, International Confectioner, Confectioner and Baker, Bakers Review,* and *Soda Fountain,* all catering to producers and shop owners. While these publications present the confectionery and related trades in a positive light, they also document overarching consumption trends by uncovering businessmen's marketing preoccupations and concerns. Since confectioners' businesses depended on catering to customers of all tastes and classes, their trade literature describes various customer preferences, as much as they tried to influence them. The topics of articles and special features reveal the contemporary marketing strategies of these businessmen—strategies such as temperance reform, which reflected popular concerns. Tools of boosterism and professional pride, journal issues are also replete with photographs and other illustrations of window displays, store interiors, shop owners, and confections themselves. While at the time these may have served solely as promotional material, today they provide crucial primary visual evidence recording the contemporary physical universe and consumers' marketplace surroundings.

The Warshaw Collection of Business Americana at the Smithsonian Archives is a terrific resource for advertising ephemera and marketing paraphernalia for businesses both famous and obscure. Its collection, which includes trade cards, booklets, and circulars, also contains a collection of confectionery boxes and labels—the kinds of evanescent packaging that was most often tossed aside. Easily browsed, the collection contains things a researcher is unlikely to see in other repositories, and was most valuable for providing examples of actual contemporary confectionery-related material culture.

Where available, personal firsthand accounts reveal the realities of nineteenth-century American life. I found published and manuscript travel accounts, diary entries, recipe books, and scrapbooks in repositories such as the Historical Society of Pennsylvania and the Hershey Community Archives. These personal traces anchor, balance, contradict, and confirm the often conflicting and always self-interested perspectives of the prescriptive and trade literature. Where firsthand accounts were not available, I turned to popular fiction, allowing contemporary writers to capture the essence of the time, trusting their ability to describe slices of reality. In addition, my reliance on existing material culture as an important informational source should be evident from the number of illustrations found accompanying the text. No mere window dressings to enliven the words, they are pieces of evidence used to bolster my argument, as crucial as contemporary textual quotations. While it is difficult, if not impossible, to know for certain how people interpreted various elements of popular culture, the amount of ephemera and advertising that still exists suggests what the nineteenth-century material universe must have been like. The sheer quantity and repetition of printed images created a gestalt, as this material culture reveals.

SECONDARY SOURCES

Of course, Sidney Mintz's book *Sweetness and Power* (New York: Penguin Books, 1985) proved highly influential and inspiring during my entire research and writing process. More than Noël Deerr's comprehensive *History of Sugar* (London: Chapman & Hall, 1949, 1950), Mintz's work considers the impact of sugar on modern political economies and places the commodity in cultural context. In addition, *Sweetness and Power* discusses both production and consumption. Mintz considers top-down and bottom-up factors in his exploration of the evolution of both the modern sugar market and sweet tooth. Mintz's work continues to develop, and I must also acknowledge his article "Sweet, Salt, and the Language of Love," *MLN* 106 (1991): 852–60, which informed the postscript in particular by pointing out and confirming to me certain salient points about sweetness in our vocabulary.

I have used Mintz as both an inspiration and a springboard—a jumping-off point from which to tell another part of sugar's story, with a different emphasis. While Mintz focuses on sugar in British culture and prefers production considerations, I have focused on the development of Americans' use of sweetness physically and symbolically. Mintz touches on sugar's symbolism in British culture; in contrast, I have made sugar symbolism in America my main focus. When I began this project, only a smattering of studies seriously considered confections themselves and the ways they conveyed meanings. Scholars' interests have changed, and I keep adding new sources to my list. The most notable works include Gail Cooper's "Love, War, and Chocolate," in *His and Hers: Gender, Consumption, and Technology,* ed. Roger Horowitz and Arwen Mohun (Charlottesville: University Press of Virginia, 1998), 67–94; Kim Hall's "Culinary Spaces, Colonial Spaces: The Gendering of Sugar in the Seventeenth Century," in *Feminist Readings of Early Modern Culture: Emerging Subjects,* ed. Valerie Traub et al. (Cambridge: Cambridge University Press, 1996), 168–90; Diane Barthel's "Modernism and Marketing: The Chocolate Box Revisited," *Theory, Culture and Society* 6 (1989): 429–38; and Susan J. Terrio's *Crafting the Culture and History of French Chocolate* (Berkeley: University of California Press, 2000). I need also mention here the important anthology *Sweetness,* ed. John Dobbing (London: Springer-Verlag, 1987), a compilation of studies on sweetness in various contexts, from its physiological effects on athletes to how it is utilized in marketing. The most valuable of these to my study was Claude Fischler's "Attitudes Towards Sugar and Sweetness in Historical and Social Perspective" (83–98).

Methodologically, works incorporating an interdisciplinary approach and focusing on issues in nineteenth-century America have guided me. These include T. J. Jackson Lears's *No Place of Grace: Antimodernism and the Transformation of American Culture, 1880–1920* (New York: Pantheon Books, 1981) for its meaningful insights into the nineteenth-century American psyche. Lears's similarly compelling study of the rise of American advertising, *Fables of Abundance: A Cultural History of Advertising in America* (New York: Basic Books, 1994), is a sweeping narration of America's burgeoning commercial culture and its influential institutions and individuals. In the same vein are Leigh Eric Schmidt's *Consumer Rites: The Buying and Selling of American Holidays* (Princeton: Princeton University Press, 1995) and William Leach's *Land of Desire: Merchants, Power, and the Rise of a New American Culture* (New York: Vintage Books, 1994), both of which chronicle nineteenth-century consumer culture and its impact on American consumers. These last three books in approach and execution—including their generous inclusion of and reliance on illustrations as primary evidence—have been highly influential to my own work and supremely helpful in understanding aspects of nineteenth-century consumerism in its larger contexts.

Although I consider this a book more about consumption and commodity culture than about food history, the work of food historians has been valuable to my understanding of sugar as a food and its place in emerging American dietary patterns. The most useful of these for my purposes have been Louise Belden's *Festive Tradition, Table Decoration and Desserts in America, 1650–1900* (New York: Norton, 1983) and Laura Shapiro's *Perfection Salad: Women and Cooking at the Turn of the Century* (New York: Farrar, Straus & Giroux, 1986; reprint, 1995), works that specifically discuss the rise of dessert culture. Harvey Levenstein's *Revolution at the Table: The Transformation of the American Diet*

(New York: Oxford University Press, 1988) incisively chronicles the transformation of the American diet during industrialization, while John F. Kasson describes how Americans ate that changing food in *Civilizing the Machine: Technology and Republican Values in America, 1776–1900* (New York: Grossman, 1976). C. Anne Wilson, in personal articles and through her edited compilations like *The Appetite and the Eye: Visual Aspects of Food and Its Presentation Within Their Historical Context* (Edinburgh: Edinburgh University Press, 1991), has explored the more elusive symbolic meanings that food engenders. The work of Susan Strasser, Ruth Schwartz Cowan, Karen Halttunen, and many others has pointed the way toward placing sugar and desserts in their proper and broader feminine domestic contexts: respectively, *Never Done: A History of American Housework* (New York, Henry Holt, 2000) and *Satisfaction Guaranteed: The American Making of the Mass Market* (New York: Pantheon Books, 1989); *More Work for Mother: The Ironies of Household Technology from the Open Hearth to the Microwave* (New York: Basic Books, 1983); and *Confidence Men and Painted Women: A Study of Middle-Class Culture in America, 1830–1870* (New Haven: Yale University Press, 1982).

Studies with more anthropological approaches have also illuminated this book. Four key figures assisted me in understanding the deep histories and meanings of what I refer to as the "confectionery triumvirate." Sidney Mintz, of course, provided key details and insights about the production and consumption sides of refined sugar, detailing everything from the cultivation and processing of sugarcane to the marshaling of exploited and slave labor to its sale in the European marketplace. Michael Coe and Sophie D. Coe provide a regionally and chronologically sweeping history of the cacao bean in their *The True History of Chocolate* (New York: Thames & Hudson, 1996). A highly accessible study that tells the story of chocolate in the premodern period, it confirms my sense that chocolate started out as a substance associated with masculinity and has helped me trace how early ideas of chocolate's powers in aiding sexual potency were perpetuated through time. Elizabeth David's *Harvest of the Cold Months: The Social History of Ice and Ices* (New York: Viking, 1994) has proven resourceful for describing the origins of ices, ice creams, and ice processing. Also relevant here is Simon Charsley's *Wedding Cakes and Cultural History* (New York: Routledge, 1992), the only work I know of which unpacks the origins and meanings of the modern wedding cake. Without his work, especially on the British history of the wedding cake, my chapter on ornamental cakes would have been sorely lacking.

Because my book is actually a study of people's relationships to their material universe, current theories of material culture and consumption informed the way I looked at and interpreted the meanings of confectionery. That people can be active consumers within commodity capitalism comes from the work of Daniel Miller, *Material Culture and Mass Consumption* (New York: Blackwell, 1994). And John Fiske, who has written widely on popular culture, including *Reading the Popular* (New York: Routledge, 1991), believes in a fluid relationship between the buying public and the popular culture around them. Together, their work has provided me with a producer/consumer framework that helps one understand how the public responded to refined sugar and why. Thorstein Veblen's classic *The Theory of the Leisure Class* (1899; New York: Dover, 1994), contemporary to much of what I describe, gives a sardonically detailed view of the consumer ethic for the American middle class and undergirds many sections of this book.

The studies of other scholars whose central focus is material culture are also important. Kenneth Ames's *Death in the Dining Room and Other Tales of Victorian Culture* (Philadelphia: Temple University Press, 1992) elucidates just how important objects were to Victorian Americans and how even prosaic things like needlework samplers held great significance for the people who made and looked at them. Adrian Forty's *Objects of Desire: A History of Commodity Design* (New York: Pantheon Books, 1986) traces the design evolution of certain domestic, occupational, and industrial artifacts. His study records the dialogue between animate and inanimate, showing the synergistic relationship between people and objects. Other similar approaches that consider the gendering of material culture specifically are two important compilations, *The Sex of Things: Gender and Consumption in Historical Perspective,* ed. Victoria de Grazia, with Ellen Furlough (Berkeley: University of California Press, 1996), and *His and Hers: Gender, Consumption, and Technology,* ed. Roger Horowitz and Arwen Mohun (Charlottesville: University Press of Virginia, 1998). And Susan Stewart's *On Longing: Narratives of the Miniature, the Gigantic, the Souvenir, the Collection* (Baltimore: Johns Hopkins University Press, 1984) explores how people used material objects as concrete items from which emerged complex psychological worlds built around fantasy.

In many places I have emphasized the materiality of various confections and have ordered and divided my chapters based on specific candies and their respective consuming audiences. What confections looked like—which necessarily reflected the way people physically manipulated sugar—was highly influential in attracting different groups of consumers. Roland Barthes's semiotics, presented in his *Mythologies* (Paris: Éditions du Seuil, 1957), trans. Annette Lavers (New York: Hill & Wang, 1972), explains why the connections between the character of purchasers and the things they bought became so naturalized. I have ventured beyond Barthes's work to consider that of Arjun Appadurai and Igor Kopytoff. Crucial essays in Appadurai's *The Social Life of Things: Commodities in Cultural Perspective* (New York: Cambridge University Press, 1997) are Appadurai's "Introduction: Commodities and the Politics of Value" (3–63) and Kopytoff's "The Cultural Biography of Things: Commoditization as Process" (64–91). Appadurai has introduced us to the idea that material things have "social lives" that change based on the various contexts that these objects inhabit. Kopytoff refines this for his own uses, believing that things not only have social lives in a general sense but also have specific individual "biographies"—life histories that are ever evolving. I have taken these concepts, especially Kopytoff's, and applied them more liberally to confectionery. Crucial to my work, and something that comes directly from the issues that Kopytoff raises, is the conviction, not only that individual confections had biographies, but also that genres of confections had biographies that were linked with the biographies of nineteenth-century Americans. This process of linking object to person during specific life stages is also what led to confections being seen as feminine entities.

Taken together, all of these components—primary and secondary sources, text and images, high and low culture—are the basis for this book.

INDEX